Structural Survey

S.L.J. Mika

ARICS, MCIOB
Lecturer, Consultant and Chartered Building Surveyor

and

S.C. Desch

QC, BCL, MA (Oxon); of Gray's Inn;
a Recorder

Second Edition

MACMILLAN

First edition (by [the late] H. E. Desch, with legal notes by
Stephen Desch) published 1970 by
THE MACMILLAN PRESS LTD
Houndmills, Basingstoke, Hampshire RG21 2XS
and London
Companies and representatives
throughout the world

ISBN 0–333–31823–4 hardcover
ISBN 0–333–31824–2 paperback

A catalogue record for this book is available
from the British Library.

Second edition 1988
Reprinted 1991, 1994

Printed in China

Contents

List of Illustrations

Preface I

When my father died in 1978 he had begun work on a second edition of this book. I persuaded a mutual colleague, Douglas Harrington, to undertake the work instead, but sadly he had not long embarked on it when he too died in harness. Happily, further enquiries led to Leonard Moseley at the College of Estate Management and finally to Reading University where Stephen Mika kindly agreed to accept the task. They have brought to it the resources of these institutions as well as their own considerable talents. On their behalf I also gratefully acknowledge help given by Peter Sealey.

The resulting book no longer bears the marks of my father's personal and particular interests in timber and trees, but it is hoped that as a result it may appeal to a wider audience whose members are otherwise essentially the same as those addressed in the first edition.

The scope of my notes on legal aspects is as limited as before. In these busy and fast-moving times the caveat at the start of them is especially important.

This illustrates a wider problem. The independent professions cannot hope or deserve to survive into the 21st Century unless they give an excellent service. The complications of modern building technology and the financial restraints of commercial and private clients make it increasingly difficult for surveyors to provide such a service. There is no chance of them doing so unless they attract new recruits of the highest ability who are prepared to dedicate most of their waking hours to their work, with their eyes on long-term achievement rather than short-term gain. My father believed passionately in the old aphorism, 'What's worth doing is worth doing well'. If this book converts others to the implementation of that view it will have served its object.

Temple, London.   Stephen Desch

Preface II

When I was approached by Macmillan Education to contribute to this revised edition it was with some trepidation for not only had Desch senior and Douglas Harrington sadly passed away, but Leonard Moseley was taken ill during the revision and was unable to complete the work. However, undaunted I agreed to take the reins and decided to start afresh with the original manuscript.

In this second edition I have restructured the book into what I hope is a simpler and more logical layout for its readers. The chapters do not follow the survey procedure itself but are loosely based upon individual elements and their interrelationship. I have retained a limited amount of the original book but have rewritten, updated and rationalised the text. I have added new material and included a number of additional illustrations. I hope readers will learn a little from this revision, and, if criticism can be made of its technical content, then this must fall upon my head alone.

I did not have the privilege of meeting Desch senior but having read his work over and over, I have come to respect the skilled surveyor he must have been. If he has been looking over my shoulder while I have been revising and extending his book, then I hope the changes I have made have met with his approval.

To Leonard Moseley I owe a great deal. His clear head and systematic approach has helped me not only in revising this publication but also, like so many others who have studied with the College of Estate Management, to pass their professional exams.

My final acknowledgements are to Philippa Sell for typing the manuscript, to Peter Dornan for his excellent drawings and to Jeanne Lapsley for her moral support.

University of Reading   Stephen L.J. Mika

Acknowledgements

Extracts from *Building Research Establishment* (BRE) *Digests* are reproduced by permission of the Controller of H.M. Stationery Office.

Permission to reproduce extracts from *Surveying Cracked Property—A Guide for Pynford Engineers* was kindly given by John F.S. Pryke MA, CEng, FICE, FIStructE, FBIM.

Permission to reproduce the extract from *Maintenance Management—a guide to good practice* was kindly given by the Chartered Institute of Building.

1 The Survey

Types of survey

The term 'structural survey' has in recent years tended to acquire a restricted meaning, being applied to surveys of properties on behalf of prospective purchasers and aimed at discovering defects or shortcomings that might influence the decision to pursue negotiations further. Those who carry out such surveys inevitably work under considerable difficulties —usually they can only observe what is visible on the surface, relying on their professional skill to deduce the significance of any symptoms. Sometimes a surveyor may be permitted to take up a floorboard here or there, lift manhole covers, probe timbers without damaging decorations and apply tests to drains, wiring, and other services. In many cases, because of the problems of making good damage caused in opening up, he cannot cut holes in ceilings to gain access to roof voids not provided with access hatches, nor can he take up flooring to either examine the joists and plates underneath or to determine the nature of the oversite concrete. Normally he cannot expect to investigate foundations, particularly if this involves breaking up concrete paving or damaging flower beds. Even where he can locate and have a sight of the deposited plans at the Local Authority's offices, there is no guarantee that the building was in fact erected strictly in accordance with the submitted details and drawings.

When floors are covered with fitted carpets and rooms are overfull with furniture, the task of the surveyor is made all the more difficult, unless he attends with expert removal men, carpet layers, joiners and the like. Most vendors would object to so large a party arriving to carry out an apparently simple operation and an average prospective purchaser would not be prepared to incur the cost.

Thus a structural survey is really an exercise in intelligent guesswork, carried out by someone able to interpret the significance of what is visible. The value of structural surveys of this kind obviously depends on the skill and thoroughness of the surveyor, and on his ability to set down his observations and their significance in a carefully written report. Such 'surveys' inevitably fall short of the definition of the word, as given

1

in the *Shorter Oxford English Dictionary*: "The act of viewing, examining, or inspecting [a property] in detail."

For certain purposes it is essential to carry out a more detailed structural survey. The objective of such a survey goes much further than discovering the existence of defects that should be reflected in the purchase price. It could involve determining the load-bearing capacity of floors and walls and the practicability of adapting the property to a different use, for example changing domestic accommodation to office use and the use of sophisticated equipment.

The two kinds of survey are so different that it is desirable to adopt distinctive terms— the popularly termed structural survey could be called a *Reconnaissance Survey*, and the true structural survey referred to as a *Detailed Structural Survey*.

Detailed structural surveys are naturally much more elaborate than reconnaissance surveys and the surveyor may require a set of measured drawings, at least of each floor and the roof. These should be true to scale and conveniently prepared on translucent paper so that the plan of any one floor can be superimposed on the plans of others. In addition to technical assistants, the surveyor may require carpenters and labourers to do the necessary opening up of floors and the like. A programme of work is very important, otherwise the surveyor will lose much time in waiting for flooring to be taken up and for holes in ceilings to be cut. It is advisable to make a preliminary inspection of the whole building, perhaps with a competent building contractor's foreman, to decide how much opening up will be necessary and when this work should be commenced. Ideally there should be three teams— one working ahead of the surveyor, one with him and a third team reinstating the floors, ceilings, etc., as the survey is progressing. Such detailed surveys can only satisfactorily be carried out in unfurnished buildings but they take time and they are costly.

In some respects the reconnaissance survey calls for almost more skill than the detailed structural survey, for the surveyor is constantly having to draw inferences that he cannot confirm by direct observation. Concentration is vital and neither the client nor the vendor should be permitted to accompany the surveyor.

In accepting instructions to carry out a reconnaissance survey, the surveyor renders himself liable in negligence to his client should he unreasonably fail to discover and report on defects. It may be no defence to a court action to have stated in the report that no guarantee can be given that there is no fungal decay or serious beetle infestation. The aggrieved client may seek to show that even in a superficial survey there were matters that ought to have put the reasonable surveyor, and not only an expert, on the alert. The surveyor could be liable in negligence if he failed to recognise and report on stained brickwork behind downpipes, ground levels outside above floor levels inside and blocked or an insufficient number of air bricks, should the existence of timber decay come to light subsequently. Areas of patched flooring, air bricks of more than one pattern, areas of recent repointing and evidence of plaster cracks having been filled are other matters that obviously should be reported. If however, the surveyor, in good faith draws the wrong conclusions from the evidence he has considered, he has not necessarily been negligent and so may not be liable.

Procedure

The principles in structural surveying are the same, whichever type of survey is required. They should follow a definite programme, probably commencing with the outside of the building and its boundaries and then working from the roof down to the foundations or vice versa. In the room by room inspection the same sequence should always be followed to ensure that nothing is overlooked—for example, ceilings, cornices, friezes, walls, windows, doors and door furniture, floors, fireplaces, electrical and other fittings.

As already stated, surveys should follow a definite programme or pattern and a number of publications are available which can help the surveyor when determining a sequence for a survey. One publication series which includes procedures for surveying traditional and unusual properties is *Structural Survey* which is published quarterly by Henry Stewart Publications.

Equipment

Notebooks, tapes, a measuring rod and a scale ruler are basic items of equipment. A piece of hardboard, probably A4 size, with a strong clip or a block of squared paper is convenient for sketching. A suitable tool for probing timber is useful, a small electrician's screwdriver is particularly suitable for testing for fungal decay because it is not too sharp, an ordinary pen-knife is dangerous as the blade tends to shut too easily. A really sharp knife or razor blade is required for cleaning up any cross section of a splinter of wood if identification with a pocket lens (say x15 magnification) is to be undertaken. The lens may also be needed for identifying beetle frass (bore dust) and for distinguishing the crystalline structure of efflorescence on plaster from fungal or mould growths. A pocket compass to fix notional compass points is much to be preferred to describing rooms as right or left, front or rear. A length of wire for testing that the vents in air bricks are unobstructed is useful, for it should never be accepted that air bricks are functional since they may be rendered virtually useless by wall plates or joists on the inside face of the wall blocking the air bricks. When inspecting two, three and four storey older dwellings, a plumb line will be essential for checking the vertical alignment of walls. A builder's 'one metre' spirit level is vital on some occasions, particularly when ascertaining the amount of fall in gutters and the falls on flat roofs. A camera is also useful to record and reinforce the observations made on site.

Overalls or a boiler suit are necessary and a breathing mask to wear in roof voids or when crawling under floors is really essential. Binoculars can be a useful aid, but they are not a substitute for close inspection where this is possible. A small mirror, about 100 x 100 mm will be useful to have sight of the conditions that cannot be observed directly.

Where electricity is available a wandering lead with a 100 or 150 watt protected lamp is preferable to a torch and 50 metres of cable should suffice for most occasions. The cable and connections must be maintained in first-class order; the bulb should be housed in a wire cage and a spare one always carried. The lamp should never be put down in a roof void, because a bulb gives out a surprising amount of heat when in use, and the risk of fire is always present.

Some surveyors like to carry ladders, a range of tools for lifting floorboards and manhole covers, equipment for testing drains and the like, but there are many advantages in having a local builder in attendance for major operations.

Fibre optic equipment is now available to provide inspection without the need to cut away and expose hidden detail. This equipment is portable and can be combined with photography and display screens to provide clear pictures of defects normally hidden from view. Clearly the use of this equipment can save much inspection time and give visual access to areas which otherwise might never have been reached. Typical situations that lend themselves to the use of such fibre optic equipment include the following:

(a) Inspection of traditional cavity walls with problems of wall tie failure, missing or broken damp-proof courses or cavity trays.
(b) Inspection of cavity fixings behind wall coverings.
(c) Identification of retaining cramps to cladding panels.
(d) Inspection of inaccessible areas under floors or behind panels where dry rot or beetle infestation is suspected.

Going a stage beyond this, it is now possible to see beyond the visible and to take advantage of the fact that everything emits infra-red radiation. Equipment is available that will detect and display on a television screen a continuous thermal picture. On the display, black sections are those which are coldest and the progression to light sections indicates the increasingly warm to hot areas. Consequently the camera and the technique which is known as thermography, detect heat loss and inconsistencies or flaws which allow such heat flow and therefore loss. Equipment is operated by specialists who can work in a wide range of situations including using the equipment from the air. Photographic records can be made of display scenes. The association of thermography with energy surveys, insulation performance, the investigation of underground mains and so on has a special and useful application to buildings.

A copy of a county street atlas such as the Geographers' *A to Z Atlas for London*, together with Ordnance Survey maps for local areas, will save much time in locating the property to be inspected and a Geological Survey map of the area could be most useful where foundation problems are suspected.

Electronic moisture meters are used for testing the moisture content of timber and walls and, provided their limitations are appreciated, they can often be of considerable assistance to the surveyor. When determining the moisture content of timber it is important to remember that the instrument is measuring the electrical conductivity of wood, which is a straight line relationship within the range of 6–24 per cent moisture content. Moreover, resistance increases with a falling temperature and decreases with a rising one, also the resistance for any given moisture content and temperature is not constant for different species of timber. For precise determinations it is therefore necessary to have the instrument calibrated for a particular temperature and species. In addition, unless the instrument is fitted with deep-probe electrodes, the moisture content reading is that of the surface of the timber, which may be appreciably below the mean moisture content of the whole piece. False readings may also be obtained where foil-backed plasterboard or where sheet foil has been used in damp-proofing repairs. Walls which contain salts and high carbon breeze blocks will give high readings as will timber which has been tanilised and walls which are badly affected by condensation. For these situations deep probe and hygrometer type meters are required, which do not give instantaneous readings, but must be left in position for periods of time depending on the particular type used. In most cases however, the surveyor should use the electronic moisture meter as an early indicator of potential problems which will require highlighting in the report. These early indications may well lead to more comprehensive testing with sophisticated equipment, at a later date.

Specimen tubes for collecting beetles or samples of frass, polythene bags for soil samples, tie-on labels for root samples, envelopes for samples of plaster or wallpaper suspected of being salt-contaminated may also be essential on occasions.

In general, splitting the work between two or more surveyors is not recommended, because it is often necessary to follow up some point that has attracted attention from room to room or floor to floor. There is less objection to using an assistant for specific tasks such as probing stiles, mullions and bottom rails or windows for 'wet rot', noting that casement stays, fasteners and the catches of windows are in good order, checking door furniture, and, externally, checking boundaries, fences and the lay-out of drains.

Instructions from the client

It is essential to obtain proper instructions, particularly for reconnaissance surveys where these are often initiated by no more than a telephone call, which must always be confirmed in writing. An adequate reconnaissance survey of even a small semi-detached house can take several hours on site even with some assistance in moving furniture, emptying cupboards under stairs and the like. Often a surveyor may be asked to advise about forming a second bathroom or the feasibility of making certain alterations, which can pose difficult structural problems and add appreciably to the time required for the survey.

For a small house, a report is likely to be a document of some 5000–7000 words. The type of report that gives little more than the usual sale particulars is worthless to a prospective purchaser, who will presumably have already checked such details for himself. Because the surveyor cannot anticipate the complications it is unwise to undertake reconnaissance surveys for an inclusive fee. A surveyor's reputation is his most important asset and the only way to preserve it is to carry out each job thoroughly.

In fairness to the client, the surveyor should advise deferring carrying out the reconnaissance survey until the purchaser and vendor have agreed terms 'subject to survey and contract' and the purchaser knows he can make the necessary financial arrangements. A 'quick survey' is only justified when the surveyor undertakes to abandon it if, early on, he finds such serious defects that he has no alternative but to advise against purchase on any terms. A surveyor should be able to arrive at such an assessment within a short time of being on site, when it is not fair to the client to complete the reconnaissance survey at possibly some considerable cost to the client. Clients seldom realise that virtually no house ten or more years of age is likely to be ready to move into without some remedial work; even in quite small houses expenditure may have to be incurred to put the property into first-class order.

Background to the survey

It is essential that the purpose of any inspection should be understood by the surveyor before he visits the site, but whatever this may be, when actually carrying out his inspection he will be considering:

(a) The initial architectural design—complicated geometry is often an indication of problems.
(b) Any subsequent conversion or adaptation.
(c) The use for which the building was erected and any change of use since that time.
(d) The initial constructional detailing.
(e) The original specification and that of any additional works.
(f) The initial standard of workmanship and possibly the supervision experienced.
(g) The standard and adequacy of maintenance that has been carried out.

(h) How the building is presently being used—for example the volume occupied by goods or furniture and the effect of this on say a heating or ventilating installation.

(i) Surrounding activities that may become hazardous—air-borne chemicals which will attack materials or rain in conjunction with combustion gases, producing sulphuric acid.

These factors will all have influence on the standard of performance of both materials and elements and of course some are more easily observed and established than others. It must be remembered that every building has performance limits.

Practice notes

Essential reading for any surveyor when carrying out building surveying work are the many practice notes published by the various professional institutions. Of particular merit are those practice notes produced by the Royal Institution of Chartered Surveyors, such as *Structural Surveys of Residential Property*. This document is essential reading for it gives excellent advice on:

(a) Taking Instructions.
(b) Preparing for the Survey.
(c) The Inspection.
(d) The Report—General Guidance.
(e) Contents of the Report.
(f) Leasehold Residential Property.
(g) Fees and Charges.
(h) Professional Indemnity Charges.

Form surveys

In the last ten years a number of forms have been produced to help the surveyor when carrying out his survey. Unfortunately the inflexible nature of a number of these forms has hindered rather than helped many experienced surveyors. It is well to remember that buildings vary considerably, unlike the standard form with its set format. Surveyors can, and often do, rely too heavily on such documentation when walking around a property, and this can lead to defects in the unusual building being missed. Standard format survey forms are useful however, and do have a part to play provided the surveyor realises their limitations. Two useful survey forms prepared by the Royal Institution of Chartered Surveyors are the *House Buyers Report and Valuation* and *Flat Buyers Report and Valuation*. These forms are not intended to be used for the very large period, highly complex dwelling or for the extremely dilapidated house, but are intended as an alternative type of report to the full structural survey for less than unique properties. The Institution's intention in producing these two standard report documents was to create a format which set out precisely the terms of reference with which the surveyor contracted with his client to supply answers to a set number of questions, in other words, a precise framework within which the surveyor could operate. The success of these schemes is reflected in the large sales of the forms from the Institution's bookshop.

Notes

A surveyor can either make his own notes or record them on a tape recorder. Provided the secretary who deals with the recording is very experienced there is much to be said for dictation, which allows the surveyor to concentrate on his observations. A sequence should be developed and followed throughout the inspection and such a system will help to eliminate failure to observe and record all the points that are, or might become, very important.

An ability to draw is a useful skill for the surveyor who undertakes large numbers of investigations. Various methods of recording observations can be used and figure 13((i) and (ii)), illustrates one method advocated by John Pryke, a consulting engineer with considerable experience in the inspection and repair of buildings. These sketches show cracks which are recorded in three dimensions to give a true impression of the defects. Figures 3 and 4 illustrate other sketch approaches which show various observations and areas for further investigation.

Measuring building plots

It may be necessary to prepare line drawings of the plot to form details to be included as an appendix. In principle, the work involved in measuring buildings and building plots does not differ from that involved in simple land surveying, but different equipment is used and different methods of booking the measurements are adopted.

A sketching block or a notebook of squared paper is preferable to the ordinary double-ruled field book, which is too narrow to be convenient for the sketch plans of buildings. The measurements are taken usually with a tape and rod. The tapes may be of steel or linen. Steel tapes are more accurate than linen and should be used on precise work, but in most cases linen tapes will suffice. It is useful to have two tapes, one 50 metres long, so that long dimensions can be obtained conveniently, and the other, 20 metres in length and used for measurements from the main tape line and for all other dimensions of moderate length. For short dimensions a rod is more useful as only one person is needed to hold it. Indeed it is an advantage to have two such rods, one kept by the surveyor and the other by his assistant so that dimensions can be taken by either of them. Electronic equipment of pocket size is available for taking dimensions, but its use is limited, in some cases by the purpose and degree of accuracy of the measurement required and in others by the many obstructions encountered.

Just as the planning of the survey lines in a field or on an estate will depend on its shape and the positions of obstructions, so the scheme of measurement of a building site will depend upon its shape and the position of the house or other buildings upon it. If a block plan is required one could start by making a sketch of the site. A base line is then chosen, such as a long straight wall or fence, which possibly could be the wall or fence along the road frontage, see figure 1(i). Usually however, the frontage of a building plot is much less than its depth, so that in most cases one of the side fences will be preferable.

The base line will be measured with the long tape, chalk marks being made on the fence or wall at all points from which long diagonal measurements are to be taken to other fences, and the distances to these marks booked. Ties to corners of buildings will be taken with the short tape, while the long tape is lying on the ground in position. Other fences, if straight, will correspond to the main survey lines of a chain survey, sufficient measurements being taken from the base line to enable them to be plotted. Offsets or ties will be taken

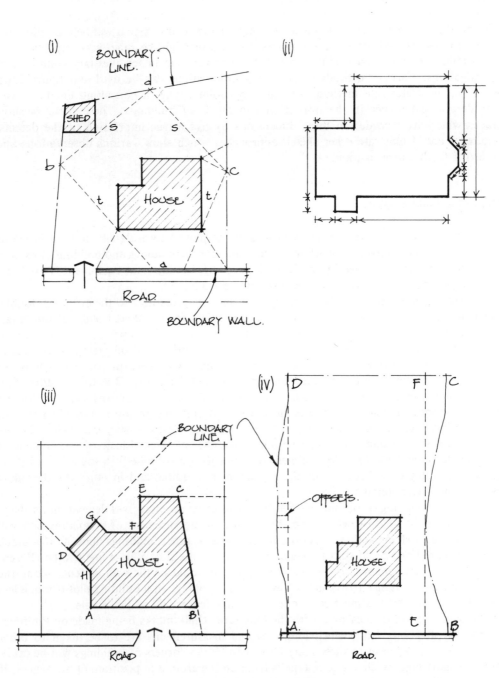

Figure 1   Site survey

from them to any nearby corners of buildings, trees and other features. When all the site measurements have been taken and an adequate number of corners of the building 'tied' or 'offsetted', then the length of each wall of the building will be measured.

A few examples are illustrated of the various problems which can arise. Figure 1(i) shows a site of large frontage with a low boundary wall between it and the road. A suitable base line is the inside face of this wall, though the thickness of the wall will need to be measured as the outside face will be the boundary of the plot. The frontage can be measured and ties taken to the two front corners of the house, two ties to each corner from noted points as shown by dotted lines. Chalk marks are made on the wall at suitable points for the long ties to the side walls, and the distances of these marks along the base line are read and noted.

The side boundaries 'b' and 'c' are then measured and the ties 't' to them from the chalk marks made on the base line. If when measuring 'b', it is not possible to get behind the shed to the corner of the plot, the measurement must be made up by dimensions taken inside the shed, making due allowance for the thickness of its walls. When measuring 'c', ties will be taken to the back corner of the house in case the building is not quite square and also as a check on the other measurements taken to the building.

The length of the back boundary 'd' is measured and this will be a check on the correct taping and plotting of sides 'a', 'b' and 'c' and the two ties 't'. Because of the shortness of the ties 't', additional ties 's', across the rear corners, may be taken as a further check. Wherever possible, the diagonal ties across the corners are so arranged that two of them are from the same point, for instance, the two ties 's' meet at a point on the rear boundary 'd'. Although this is not essential, it is less likely to cause confusion of chalk marks on the walls, and fewer number of distances need to be recorded. In two cases ties have been arranged to touch the corners of the buildings as an additional check to the position of the building. Diagonal ties, such as 's' and 't', should be as long as is reasonably convenient, but generally should not exceed that of the type with which they will be measured.

If the corners of the building are square (and buildings are much less likely to be out of square than are building plots) all that is now needed is to measure all the wall projections (porches and bays) and recesses. In order to have space for these dimensions it is usual to put these on another sketch, showing more detail and sketched to a larger scale than that used for the plot, figure 1(ii). The shed is dealt with similarly, but as it is out of square, the lengths of its four sides and its two diagonals are measured. These measurements will be taken internally (since the diagonals could not be measured externally) and the thickness of the sides noted.

Figure 1(iii) shows a method which can be adopted for buildings whose corners are not square. This difficulty could be overcome by taping every angle of the building by ties to points on the boundary walls, but a very large number of measurements would be needed. If the corners 'A', 'B', 'C' and 'D' are tied from the boundaries and corners 'A', 'E' and 'F' are square, then the points at which the continuation of 'EC' intersects the east boundary, 'DG' intersects the north boundary, 'HD' intersects the west boundary and 'CB' intersects the south boundary, are noted and marked with chalk on the boundary lines before the lengths of the latter are measured. Whenever a chalk mark is made on a wall or in the ground it should be done in a distinctive way, usually in the form of an arrowhead, the point of the arrow being the precise point at which the measurement is taken.

Sometimes the boundary fences of a building plot are not straight, in which case some modification is needed to the methods described and figure 1(iv) shows such a case. The western boundary can be dealt with by taping from corner 'A' to corner 'D' and taking offsets to the fence. The tape cannot be laid down in a straight line from corner to corner

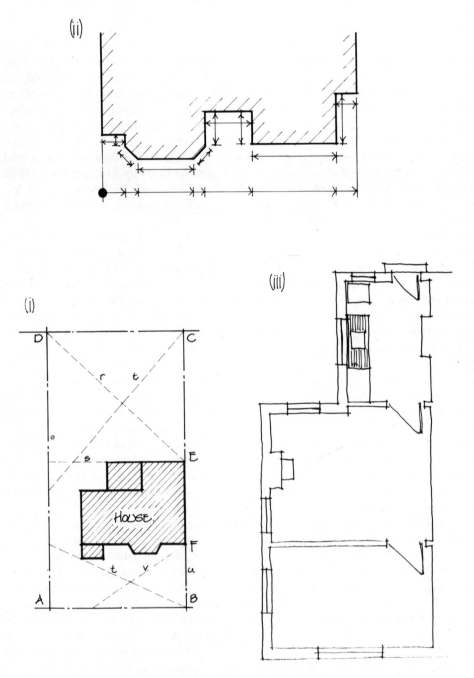

Figure 2 Measuring buildings

on the eastern boundary, but must be put in some such position as 'EF', and again offsets taken from it to the fence. Taking offsets from a tape acting as a base line tends to introduce errors owing to the difficulty of keeping the tape straight and there may be occasions where the use of a chain will be preferable.

Figure 2(i) shows a typical case of a block plan for a semi-detached house. The long western boundary fence 'AD' could be the base line and the corners 'B' and 'C' tied from it by the measurements of the front and back walls and the two ties 't'. The corner 'E' of the house is tied from the base by the measurements 'r' and 's', and the corner 'F' from the front plot boundary by measurements 'u' and 'v'. The principal corners of the house would be fixed by ties or offsets from the most convenient fence lines and the usual detailed measurements of the house taken.

Measuring buildings

When measuring long walls, broken by details such as doors, windows, bays, porches etc., there are two methods of procedure. One is to take each straight length independently, as shown by the dimension lines recorded close against the building in figure 2(ii), and the other is to take 'running measurements', all read from one position at the zero end of the tape, as shown by the bottom recording. The latter method is the quicker and it avoids the 'cumulative error' which may occur when a number of dimensions are added together. If however, there are big projections from the main face of the building, so that in order to take running measurements it would be necessary to hold the tape at a considerable distance from the main face, the method is not so reliable. A better method is to take separate measurements of each straight length and to check their total at the time of measurement by an overall dimension. When plotting the work the overall dimension should be set out first and then divided into parts from the separate measurements.

In order to prepare a block plan of the site of a building, the whole of the measuring can be done outside the building, but in the case of a terraced house, particularly if it is out of square, many of the measurements must be taken indoors. Even in this case however, it is only the principal dimensions of the rooms and the thicknesses of the walls which have been ascertained, for the objective is the preparation of a plan of the site and an outline of the building in its proper position upon the site. It may be however, that the client requires an accurate plan of the interior, either of the ground floor or any other floor. All the details of the interior on that floor, such as doors, windows, chimney breasts, fixed cupboards, etc., have to be measured and shown on the plan. On the other hand, it will not be necessary to show garden, yard, boundary fences, etc., consequently, all the measuring is carried out indoors, except that the exterior measurements of the walls should also be taken, so far as is possible.

The first thing to be done is to go over each floor making a rough sketch plan of the rooms, corridors and staircase, with the walls being shown merely by single lines. No dimensions need be taken at this stage and this will act merely as a key or guide in preparing the more detailed sketch or sketches, on which the dimensions are written, see figure 2(iii).

The detailed sketches need to be drawn large enough to contain all the necessary dimensions without crowding the figures. If the building is a small one and the sketch and the figures are made very neatly, it may be that all the rooms can be contained on a single detailed sketch. In most cases it will be better to have a number of detailed sketches, each containing a single room or a group of rooms and the key plan will show how they all fit

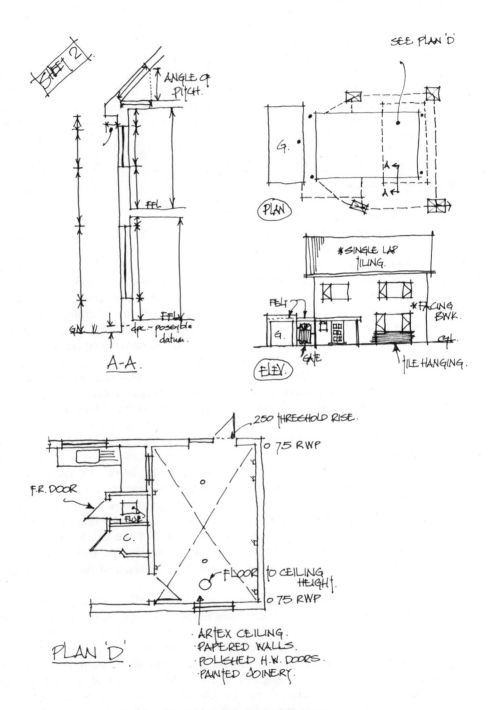

Figure 3 Typical notebook sketches

together. If it is decided to include all the building in one detailed sketch it need not be sketched out immediately. One could start with one long wall and add the rooms one by one as they are measured. A good draughtsman will be able to sketch the whole layout freehand, without using a straight-edge or set-square, and yet manage to get the rooms in good proportion. One less skilled will be wise to rule the walls and sketch the rooms roughly to scale, and probably the most convenient method of achieving this result without loss of time is to do the work on a sketch block with squared paper. The 'scale' can be made any convenient number of squares according to the size of the house and the sketch block.

A surveyor should never be satisfied with only the length and breadth of a room. All four walls and the two diagonals should be measured. If the opposite walls have equal length and the diagonals are also equal, the length of the diagonals need not be recorded, and only one dimension of length and one of breadth need to be noted, for in these circumstances the room is rectangular. If however, the opposite walls or the diagonals are not equal, all these dimensions must be entered, for one diagonal will be needed to enable the room to be plotted and the other to enable the work to be checked.

The thicknesses of walls and partitions, positions and widths of doors and windows, widths of chimney breasts and depths of recesses on either side, are items which must not be missed. Both sides of all projections and recesses should be measured, for the thickness of a wall will sometimes change at such points and the eye is very liable to be deceived. If both sides have the same projection only one dimension need be booked. Fixed cupboards are usually shown, along with items such as lifts, serving hatches, etc., but trivial fittings may be ignored.

It is always advisable when circumstances permit, to take overall dimensions right through the building and to check these with the sum of the separate measurements and wall thicknesses before leaving the site. In the case of a detached building these overall dimensions can be taken on the outside of the house, but in the case of a terraced house, the surveyor must do the best he can indoors. If a complete overall dimension is impossible, it may still be possible to measure some fairly long lines through the open doors of continuous rooms, along corridors and so on.

Some typical notebook sketches are shown in figure 3.

2 Building Decay

Change of appearance

The change of appearance, as well as the behaviour and performance, of materials, components and elements, has always been significant when examining buildings. This consideration may have been an important factor at the design and specification stage of a building and it must also be taken into account when inspecting and reporting on buildings. Laboratory testing of materials to predict weathering and performance is not always reliable because the testing applied to the material may in fact lead to a breakdown of the sample, when in a normal situation, this may not be the case. On the other hand, some tests are very appropriate and the surveyor would be wise to temper his observations with researched facts. There is considerable knowledge and experience of weathering to be gained from examining existing buildings; after all, examples of weathering and failures are all around us and these should be carefully observed. In addition to looking around, a surveyor should be aware of the large number of books, journals and investigative reports on materials, building components, performance and defects, which are published regularly. A number of reports produced by the Building Research Establishment (BRE) will be of particular use to the building surveyor, these include Digests, Information Papers, Defect Action Sheets, Current Papers and Technical Notes.

Factors causing weathering, and possibly therefore change of appearance, include rain, moisture and thermal movement, frost, heat, wind, atmospheric pollution and biological agencies. Contributing to these phenomena is the situation in the building (where stresses, for example, may change), inadequate design detailing (for example, no coping to a parapet wall) or sudden changes of environment. In these days it is possible to have building work, previously below ground, exposed for a comparatively long time for new road construction or some similar engineering work. When in a covered state the performance may have been perfectly adequate but in the changed situation, failure or at least change of appearance is possible.

The enclosure or envelope provided by buildings to protect man and his activities from

14

the weather is, of course, exposed to the weather and consequently subjected to regular and sometimes quite dramatic changes. Sometimes the cycle is so rapid that mechanical failure is experienced, but usually the time scale is not critical. At times, the change of appearance may be considered to be an improvement, for example, the patina forming on copper changing its colour to green. Changes, however, are part of the inevitable process of decay and eventual ruin of buildings which cannot be stopped. The process can be slowed down of course by maintenance, and to a point retarded by replacement of the failing materials with new ones.

As a reporter on decay the surveyor should be concerned about the physical, biological and chemical degradation of materials and their weathering performance when assembled, fixed, applied and so on in buildings. There may be some dangers when considering single materials in this way and in some cases, it is perhaps more sensible to consider larger assemblies or elements. For example, single bricks can be exposed for weathering tests establishing certain qualities, but to demonstrate the effect and effectiveness of damp-proof courses, mortar, pointing, copings and other design features, panels of brickwork in its various forms will need to be built.

BRE Digests 45 and *46* cover 'Design and Appearance', and although principally concerned with the effect of design on weathering and the appearance of Portland stone and other facing materials, these digests do provide a most useful background for the whole problem of appearance. As part of the problem one should consider the cleaning of external surfaces for this may be a maintenance activity which can be important, particularly when contemplating the arresting of deterioration. Clearly, if cleaning is undertaken the appearance of a building can be dramatically changed. The choice of cleaning method should have been carefully selected and undertaken and the type and condition of the surfaces must be carefully inspected before a specification is written. *BRE Digest 280* 'Cleaning external surfaces of buildings', gives the background to this activity and describes the methods that may be adopted.

It is in the external appearance that the greatest changes can be observed, but the problem arises internally as well, for example, pattern staining on ceilings. This can be caused by a number of factors such as the deposit of dirt by convection from the heating system; the normal occupation movement and activities in the use of the building, industrial processes and so on, can all have a significant effect on appearance as indeed can the cleaning processes that are employed.

Externally, the interaction of dissimilar materials, their juxtaposition, the possible presence of water soluble constituents and organic growths, all can combine or act singly with the micro-climate to produce an irregular change of appearance, even on the one-building elevation. In some materials, irregularity in surface texture and even shape are tolerable and indeed desirable, handwrought timber for example, but in others, and particularly those that are applied as finishes in a thin form, any irregularity is objectionable. Certain materials in combination also produce an appearance that is quite different to their individual appearance. For example, stone in conjunction with large areas of glass will appear very shabby even when very slightly weathered, when in fact on its own, the appearance of the stone would be considered enhanced when in this 'weathered' state.

Weathering

Another cause for change of appearance which is usually considered detrimental is the weathering away of a surface. Some materials which have the same composition throughout can lose their surface without change of appearance but with those that are finished with an applied surface, the change or loss of surface can be dramatic and disturbing.

It is clear now that we live in a state of rapid change. Building techniques also change so quickly that frequently traditional methods are abandoned in favour of the new. The experience reflected in traditional practices is forgotten or even worse, is considered old-fashioned. What had become almost second nature in building quite suddenly is left out of current techniques and it is possible that in some instances we may have to reconsider old habits and traditions. Frequently new buildings are disappointing and to a discriminating eye the signs and ingredients of change and even failure are apparent. Visible change is inherent in most building materials, after all they are continuously exposed and in some situations change can occur very quickly. It is possible in certain climates and situations for change from atmospheric pollution and rainwater washing down to be evident very early on. This can occur even before any chemical or physical change begins. It follows that the cleaner and drier geographical regions are less likely to produce buildings suffering from changed appearance than those which are 'dirtier' and wetter.

Careful designing and detailing initially will combine to reduce the effects of appearance changes and this should be coupled with some knowledge of the chemistry of the materials when drafting specifications and details. As always, careful observation of existing buildings will be a first class guide for the surveyor. The experience gained in this way will be most useful in predicting performance and appearance changes when reporting on the present condition of a building.

It should be remembered that certain changes tend to enhance rather than destroy or disturb appearance. For example, Portland stone elevations may weather naturally and take on a beauty that can be destroyed, or at least marred, when cleaning takes place.

Failure

The client as occupier or owner of the building will have certain expectations for his property and will anticipate that these expectations will be satisfied for a long time, subject perhaps to reasonable maintenance (if the client is aware of property management responsibilities). It is quite likely that the first notice of any defect is a lay one, and lay observation will likely exaggerate a defect to failure level rather than attributing it to natural decay or some external factor. It may be that the failure constructionally is trivial and therefore will involve little expense in repair and can in fact be left without attention. On the other hand, a serious fault is likely to be expensive in repair and may be disturbing if left without attention. At the extreme, the defect may be dangerous and will require temporary, followed by permanent, attention.

This anticipation of failure in performance is a matter of concern as is the frequency of failure in buildings and services. Reference to the published information in research documents, the professional journals, the press, radio and television programmes will emphasise these problems and illustrate the growing awareness of clients. Common faults are continually being reported and they include the following:

(a) moisture condensation, dampness, mould growth, water penetration;
(b) frost damage spalling, heave, thermal, moisture, structural movement;
(c) corrosion;
(d) structural collapse.

It is with this background knowledge that one should look at buildings, perhaps wondering why such a state should exist, considering all the accumulated experience possessed by those in the construction industry (especially as the construction industry in the United Kingdom is probably the most controlled industry in the world). Buildings start to deteriorate as they are being constructed and materials, elements and components may be damaged even before assembly and delivery to the site.

Materials not only react with other materials when in contact, but they also react to the environment in which they are placed. Therefore, when inspecting materials one should consider the macro, mezzo, micro and crypto-climates for the building in question and assess the performance in use. A variety of hostile environments exist in the UK which directly affect building materials—environments such as high or low temperatures, high humidities, corrosive atmospheres, vibrations and stresses, and misuse. Failure to appreciate the effect of these agencies may have resulted in a poor specification or constructional detail which often leads to a reduced performance standard. If the building being inspected is a new one, then it is a knowledge of these points that may help to predict the performance of materials or elements.

Performance and reliability

The prediction of performance or reliability can be difficult, and buildings which may have a design life of say sixty years may be removed in only a few years for reasons quite unconnected with failure. Where buildings have existed for a long period of time, some parts or components may have been renewed, possibly more than once, but this may be due to changing requirements as well as to failure or to substandard performance. However, irrespective of the difficulty, clients and prospective purchasers will often require accurate predictions from surveyors.

The following table is extracted from the book *Maintenance Management—a guide to good practice* published by the Chartered Institute of Building and gives some basis for predicting performance.

Item	Life expectancy (years)
Brickwork	
pointing first	50–60
subsequently every	25
Roofs	
slates	60
tiles	50
asphalt	20
felt	10
Electrical installations	25

These figures are indications only and many examples can be given where they are exceeded without anxiety or not achieved owing to extraneous factors.

It would be prudent for any building owner to arrange for a programme of testing and inspection to be commenced as soon as building is completed or even during the physical construction, if this is possible. Without this ideal situation, a surveyor has to be something of a detective and must apply testing procedures and performance standards that are appropriate for the particular circumstances. Testing obviously has to be non-destructive when being carried out in an inspection, and may involve elaborate equipment requiring specialists, but often decisions and opinions may be sufficient which are the result of direct visual observation, knowledge and experience. A systematic approach can be developed for all the elements of a building and an association of ideas in this connection is shown in the following list and illustrated in figure 4.

Defect or Substandard Performance

Symptoms	—	for example, a damp patch on the ceiling under a flat roof
Diagnosis	—	damaged roof covering
Investigation of cause	—	structural, thermal or moisture movement, differential movement of materials, mechanical damage, design specification, workmanship faults maintenance, sunlight moisture ingress, pollutants, biological agencies, proximity of unsuitable materials
Recommendations	—	estimate of cost and specification of remedial measures and possible improvement in design/construction.

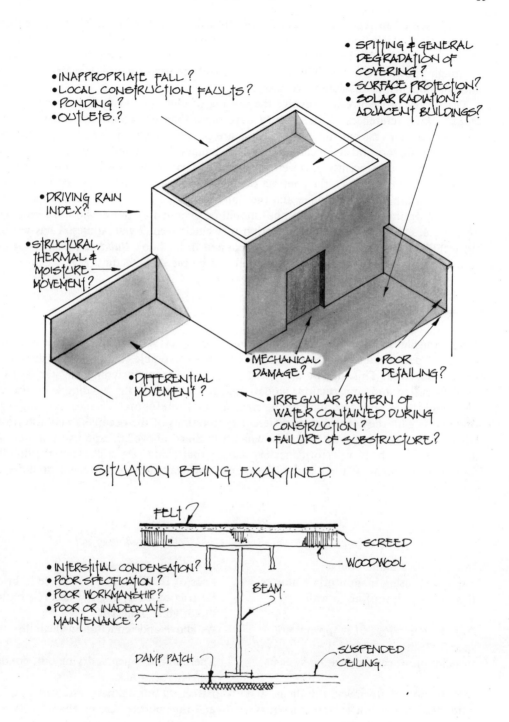

- INAPPROPRIATE FALL ?
- LOCAL CONSTRUCTION FAULTS ?
- PONDING ?
- OUTLETS ?

- SPITTING & GENERAL DEGRADATION OF COVERING ?
- SURFACE PROTECTION ?
- SOLAR RADIATION ? ADJACENT BUILDINGS ?

- DRIVING RAIN INDEX ?

- STRUCTURAL, THERMAL & MOISTURE MOVEMENT ?

- MECHANICAL DAMAGE ?

- POOR DETAILING ?

- DIFFERENTIAL MOVEMENT ?

- IRREGULAR PATTERN OF WATER CONTAINED DURING CONSTRUCTION ?
- FAILURE OF SUBSTRUCTURE ?

SITUATION BEING EXAMINED.

FELT ?

SCREED

WOODWOOL

- INTERSTITIAL CONDENSATION ?
- POOR SPECIFICATION ?
- POOR WORKMANSHIP ?
- POOR OR INADEQUATE MAINTENANCE ?

BEAM.

DAMP PATCH

SUSPENDED CEILING.

Figure 4 Site sketch of construction

It is essential to reach an assessment of any defects that is impartial and based on the investigations carried out. The assessment may depend on how much is to be investigated in the survey and the degree of accuracy that is required, but whatever these factors may be, a thorough knowledge of building construction, and the performance of materials, elements and components, is absolutely necessary.

The survey procedure will involve the seeking of clues and evidence, recording, testing and checking references. A defect may have more than one cause and these in turn can have several sources, but fundamentally we are concerned with dampness, movements due to different factors and chemical or biological changes. If the building under inspection was built comparatively recently or is well documented, then written sources of information may be available, possibly even the contract documentation. It will be necessary at times to see parts which are normally hidden and this can mean having the construction opened up or the use of optical probes. These probes should be operated by those who have training and experience, especially in the interpretation of what is seen. Lyall Addleson has written of the techniques where greater accuracy is required in his book *Building Failures—A Guide to Diagnosis, Remedy and Prevention* published by the Architectural Press.

Cause and effect

When considering older buildings, for example, the pre-1900 housing stock, it is clear that a level of performance has been achieved to some acceptable standard, in order for them to have survived. This may be attributed to several reasons including the standard of the original design and construction, modest changes in the way the building is occupied and used, and the in-built flexibility of the structure and careful maintenance. A changing pattern of use with the consequently different expectations of the occupiers may lead to work being undertaken that may upset the balance outlined above. Prospective purchasers will almost certainly have a performance expectancy that could very well affect the building if implemented. Some of these activities that are undertaken by or for clients and occupiers are set out below.

Requirement	*Possible effect*
Insertion of a damp course	Drying out, drying shrinkage
Repointing using inappropriate mortar	Loss of flexibility, efflorescence in bricks
Brushed waterproofing to wall	Restriction of drying out and respiration of the original materials
Applied renderings of inappropriate specification	As above and reduction in flexibility
Installation of central heating systems	Thermal movements, drying out, conditions for dry rot attack
Introduction of stormwater drainage and hard landscaping, for example, pavings, patios	Effect on normal moisture content of soils and thus on the bearing characteristics
Large scale site de-watering	Historic bearing capacities and ground conditions may change.
New electrical installations	Effect of notching joists

New plumbing installations	As above, and also increased loads from tanks, etc.
Insulation of roof spaces	Thermal equilibrium changed, thermal movement and condensation
Structural alterations	Weakening of the structural integrity of the building
Extensions	Differential movements incompatible materials

Contributory factors

In modern construction, among the many factors that may produce an unsatisfactory performance and which should be looked for while carrying out the inspection are:

(a) The use of concrete which is not so sympathetic to movement.
(b) The association of elements with different thermal movement characteristics.
(c) Thermal and drying shrinkage of materials.
(d) Minimum size timbers and other materials.
(e) A poor standard of initial workmanship.
(f) Condensation and the lack of ventilation.
(g) The dramatic growth of 'do-it-yourself' which may have caused damage to the construction.
(h) The effect on buildings of soil movement, particularly clay soils.
(i) The presence of asbestos products and the effect on health.
(j) The use of mastic sealants.
(k) The corrosion of metals, including rusting of reinforcement and cavity wall tie failures.
(l) Carbonation of concrete and the use of high alumina cement.
(m) Inadequate falls to flat roofs, gutters and drainage.

Two useful guides for surveyors on the typical defects found in buildings are *BRE Digest 176* 'Failure patterns and implications' and *BRE Digest 268* 'Common defects in low-rise traditional housing'.

3   Substructure

Building plans

It is rarely necessary to inspect the foundations in the course of reconnaissance survey, unless there is evidence of settlement damage, the cause of which is not obvious. Even when desirable it may not always be practicable because of the proximity of drains, paving, or flower beds. If the house was built within the last fifty years, some information may be obtained from inspection of the deposited plans at the offices of the Local Authority, although there is no guarantee that a property has been built strictly in accordance with such plans. With older properties, enquiry at the appropriate office will often yield some useful information. Since 1947 it has been necessary to submit plans and obtain planning permission for structural alterations. Not all owners have done this, and a new owner wanting to make further alterations may have to carry out additional work to earlier alterations to make these conform with statutory requirements.

For property within the former Greater London Council (GLC) boundaries, District Surveyor's offices (now Building Control under London Boroughs, the Temples and the City Corporation) will have plans of any structural work that has been carried out, such as underpinning. Similarly, major remedial work completed in the course of dry rot repairs may also be recorded. For property outside the former GLC boundaries, building control offices within Local Authorities will hold such information.

Wherever practicable, it is advisable to inspect these plans before carrying out the site reconnaissance survey. Many Authorities will not allow plans to be inspected without prior authorisation by the owner, and this should be secured in advance. With post-war houses the Local Authority may have Building Control Officer's Notes on site inspections, or the Building Control Officer for the area may, from memory, be able to fill in useful details or clear up points not readily explicable. For example, during the investigation of foundations in connection with damage from tree root action, foundations substantially deeper than normal for light domestic work in the locality were discovered. A call on the Building Control Officer provided the explanation. The foundation trenches, which were in London clay, had been dug in very wet weather, resulting in the bottom of the trenches being reduced to a slurry, and the Officer had insisted on the foundations being deepened by 300 mm (one foot). This removed the earlier supposition that the deeper foundations had been necessary because of some local defect in the load-bearing properties of the clay. On another occasion, the cause of minor cracks in an external wall could not be established. The deposited plans could not be traced because at the time, that Local Authority filed plans under the name of the original owner of the property. Exploratory digging showed that there was no strip foundation concrete under that particular wall, although the house was only about thirty years old. The purchaser was advised that minor movements might persist, and an adjustment of the purchase price was agreed with the vendor.

Foundation depth

It is important to investigate the depth of foundations of pre-war or earlier houses for it is not uncommon to find that substantial and apparently well-built houses have foundations no deeper than 375–500 mm (15–18 inches). Investigations are particularly important if there is a proposal to add to the existing structure, for such shallow foundations would not be permitted for modern extensions. These new additions would have to conform to the current Building Regulations, and bonding in the new work with its deeper foundations to existing walls with relatively shallower foundations can give rise to differential movement between the old and the new work. In addition to the problem of connecting new to old, shallow foundations on particular soils such as shrinkable clays can encourage foundation movement following prolonged dry spells. *BRE Digests 240, 241* and *242* 'Low-rise buildings on shrinkable clay soils, Parts 1, 2 and 3' elaborate on this problem.

Geological Survey Maps

It is always desirable to inspect the Geological Survey map for the area, selecting a scale that is adequate to pinpoint a single property. It should be remembered that some of the maps were published many years ago and much residential development has occurred since. Some Local Authorities have copies of the maps of their area, and some have produced their own geological maps. These may be more accurate than the official Geological Survey maps because details have been filled in from actual excavations and investigations made in the course of development of the whole area.

Soil investigations

Investigation of the soil should be an essential preliminary to any building work, and it is to be recommended when surveying a newly built house or bungalow because there may not have been time for settlement to develop. It is of course essential if the surveyor has been asked to advise on alterations and additions to existing property. The subject of soil investigation is dealt with at some length in *BRE Digests 63, 64* and *67* 'Soils and foundations, Parts 1, 2 and 3'. See also *British Standard Code of Practice, CP 2001: 1957* 'Site investigations'.

It may be argued that soil investigations are more the sphere of the architect or structural engineer developing a new site, but it is pointed out in *BRE Digest 64* that:

"It is of particular importance at the present time, when the possible use of sites that have been avoided as building land in the past is being considered. This may seem to be stating the obvious but in fact it is known that schemes have been prepared, and in some cases work actually started before any consideration has been given to below-ground conditions at the site and whether they are suitable for the project in hand. The extent of the investigation will depend largely on the situation, size and type of proposed work."

This Digest is intended to emphasise the need for an early appraisal of the site so that any special measures that may be required to deal with difficult conditions can be planned at the outset. Many other aspects which bear on the intended use of the site, and which will need to be considered, such as leases, wayleaves, limitations on use and so on are outside the scope of the Digest.

The warning cited should certainly be borne in mind when carrying out a reconnaissance survey of a house or bungalow on a new estate. The powers of Local Authorities in regard to development of sites are appreciable, but, provided the Building Regulations are observed by developers, the powers of Building Control Officers are, in practice, limited. For example, on one occasion a dispute arose between a purchaser of a new bungalow and the developer when dry rot resulted from flooding of the oversite concrete because the site had been inadequately drained. The trouble in this case was that the site had been subject to inundation before it was developed, because a hard pan existed under the sandy top soil, and below the pan was a soft clay. As a result of the development much of the land was covered with impervious materials such as the roofs of the bungalows and the tarmac roads. Deepening of an existing stream proved inadequate to cope with the large volume of water in times of heavy rain, and, with so much of the porous sandy soil being covered by the buildings and roads, there was insufficient land available to absorb surplus water on some parts of the estate. The result of this situation caused the water table to rise above safe levels and flooding of the concrete oversite had occurred. The development was completed before the advent of the guarantees that are now available on new property, and the owner had to resort to court proceedings. The developer was required to remedy the inadequate drainage and to eradicate the dry rot.

On another occasion the original object of enquiry was a row of trees close to the boundary of a new estate, but not under the control of the developer. The Geological Survey map of the area indicated that the soil was not a shrinkable clay, and hence the risk of damage from tree root action to foundations did not arise, and no drains were sufficiently near for there to be a risk of them becoming blocked. Some of the trees were, however, too close to the house the client was interested in because of a potential risk from falling branches. It seemed advisable to check with the Building Control Officer the precise nature of the soil and depth of the foundations. These enquiries revealed important matters that could not have been deduced from site observations. The site was apparently one known to be subject to inundation and the Building Control Officer had, in consequence, insisted on the foundations being taken to a depth of 1500 mm (5 ft). This ensured that settlements would not be a problem, but it meant that the gardens would be wet for long periods and lay-out would have to be restricted to what would grow in such unfavourable soil conditions. Of no less importance to the new owners were the council's plans for developing tall blocks of flats on the adjacent land, which would overshadow

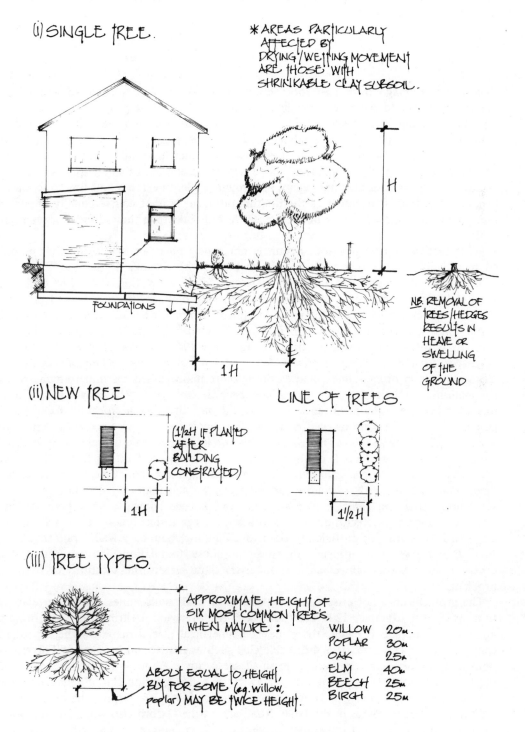

(i) SINGLE TREE.

* AREAS PARTICULARLY AFFECTED BY DRYING/WETTING MOVEMENT ARE THOSE WITH SHRINKABLE CLAY SUBSOIL.

H

FOUNDATIONS

1 H

N.B. REMOVAL OF TREES/HEDGES RESULTS IN HEAVE OR SWELLING OF THE GROUND

(ii) NEW TREE

(1½H IF PLANTED AFTER BUILDING CONSTRUCTED)

1H

LINE OF TREES.

1½H

(iii) TREE TYPES.

APPROXIMATE HEIGHT OF SIX MOST COMMON TREES, WHEN MATURE:

WILLOW	20m.
POPLAR	30m
OAK	25m
ELM	40m
BEECH	25m
BIRCH	25m

ABOUT EQUAL TO HEIGHT, BUT FOR SOME (e.g. willow, poplar) MAY BE TWICE HEIGHT.

Figure 5   Tree heights and foundations

the houses, unquestionably adversely affecting their value. This experience underlines the importance of making enquiries at the offices of the Local Authority when surveying houses on new estates.

Tree roots

The risk of damage from tree root action in shrinkable clay soils is something that the surveyor should have very much in mind when surveying properties or inspecting potential building sites in such areas. A suggestion has been that to reduce the risk of damage as far as possible, it is advisable as a rough guide, not to erect buildings on shallow foundations closer to single trees than their height at maturity, see figure 5(i). The roots of groups or rows of trees competing for water over a limited area can be more extensive than just indicated and again, as a rough guide, one and a half times the mature height of the tree is suggested as the limiting distance. It is of course equally important that young trees should not be planted closer to buildings than these distances, see figure 5 (ii).

Settlement caused by tree root action is not a new phenomenon, although it is only comparatively recently that roots have been linked with damage of this kind. The first case to reach the courts was *Butler* v. *Standard Telephone & Cables Limited* (I A.E.R. 121), which was decided in 1940; this case concerned Lombardy poplars. The Geotechnics Division of the Building Research Establishment commenced studies on the subject in about 1943, and the first Digest, *BRE Digest 3* 'House foundations on shrinkable clay soils' (now out of print) was published in February 1949. Previously it appears to have been accepted that cracking of plaster and similar minor damage to houses on clay soils was inevitable in dry summers, and anything more major called for underpinning of the foundations. Comparatively few cases have been determined by the courts and of those that have been reported in the Law Journals, the majority have been cases of damage by poplars. It is a mistake, however, to think that this is the only tree species the surveyor need consider. Elms, willows and ash can cause no less devastating damage at very considerable distances from the bole of the offending tree, see figure 5(iii). In fact all trees, and even shrubs and climbers, can cause damage. Cherry, elder, fruit trees, horse chestnut, lime, London plane, oak, sycamore and even wistaria and rambler roses, have all caused substantial settlement damage; even a hydrangea has been known to be responsible for a fracture in a wall. In effect, no tree or shrub can be regarded as innocuous in a shrinkable clay soil if it is growing sufficiently close to a building, boundary wall, drain or paving. *BRE Digest 298* 'The influence of trees on house foundations in clay soils' is an excellent text on this subject as it summarises quite clearly the solutions to tree root problems.

The problem does not arise on sandy, gravel or chalk soils, unless these are relatively thin layers, for example, 1200–1800 mm (4–6 ft), overlying a shrinkable clay. In some areas where the soil is chalk, there may be superficial glacial drifts on the top of the chalk, either of gravel or firm shrinkable clay such as that found in parts of Folkestone. Soils in many districts vary over quite small areas, for example, parts of Chelsea are on the River Terraces, whereas not very far away the soil is London clay. Similar variations are to be found in the city of Oxford; the college buildings of Corpus Christi college are on gravel, whereas houses in the Iffley Road are on the Oxford clay beds. In Hampstead there are three different formations, London clay over the greater part of the area, and

Claygate beds, capped with Bagshot sands, on higher ground. London clay has been encountered in an area where the soil was predominantly gravel, and there was substantial damage from tree root action to a block of flats. The explanation was that old gravel pits had been used for depositing clay excavated in the course of constructing one of the first underground railways.

Soil recovery and trees

The other aspect to be considered in connection with trees is soil recovery after trees have been felled. Attention is drawn to the following passage from *BRE Digest 63*:

"Swelling caused by tree removal. When trees are felled to clear a site for building, considerable time should be allowed for the clay (which was previously dried by tree roots) to regain water. Otherwise there is a serious risk that as the clay swells it will lift the building. And because the swelling is most marked close to the site of trees which have been removed, damage is likely from relative movements... The pressures that dried clays develop when reabsorbing water are often greater than those applied by shallow foundations, and the resulting, upward movements can continue for several years. For example, regular measurements made by the Building Research Station on an office block built in 1959 on a site cleared of trees a year previously, show an upward movement of about 6 mm per annum since construction was completed. Observations suggest that these movements may continue for up to ten years."

In another case, much more substantial movement had occurred in a row of single-storey cottages built over the root systems of two elms, felled to permit building operations. Parts of the walls of these cottages were lifted by as much as 125 mm (5 inches) over a period of eleven years, causing persistent cracking of the walls because soil recovery is not uniform. Such massive movement of walls in the course of soil recovery appears to be associated more with land occupied by the root systems of trees prior to buildings being erected than from movement following removal of trees whose roots have trespassed under the foundations of an existing building. There is no doubt, however, that some movement does occur in these latter circumstances if the offending trees were close to the damaged building for example, 3000–4500 mm (10–15 ft) away. Even more surprising can be the severity of damage that follows repair of defective drains or leaking mains water supplies where there are substantial trees in the vicinity. Severe fractures develop in as little as nine to twelve months after the repairs have been effected if abnormally dry weather follows. Most new developments today enjoy the benefit of an agreement and insurance policy issued under the National House-Building Scheme by the National House-Building Council (NHBC). The NHBC is an independent non-political, non-profit-making body approved by government and recognised under statute in Section 2 of the Defect Premises Act 1972. A surveyor must study the precise wording of the NHBC guarantee being offered with a property if he is to advise a client on the action to take when a defect arises on a new property.

In shrinkable clay soil areas the proximity of near-by trees is obviously most significant, but it is equally important in such areas to discover whether trees have been felled to permit the development. Not all Local Authorities are aware of the possible consequences of developing well-wooded sites in shrinkable clay areas, and they may permit traditional 'shallow' foundations in circumstances where short-bore piles should have been used. Subsequent problems arising from soil recovery after the trees have been felled may not be covered by the guarantees if these provide only for damage arising from subsidence.

In the New Towns and similar large scale development there is the additional hazard that may arise as a result of the policies of their landscape architects, when trees that are really unsuitable in the middle of urban development are retained. The position is made all the more difficult where such trees are the subject of a Tree Preservation Order. Protracted negotiations may be necessary to persuade the Authority to set aside an order once made. If the soil is a shrinkable clay a new owner would, of course, be able to fell the trees on his land that are the subject of a Preservation Order to abate a nuisance to his neighbour but, in turn, his own property might suffer damage from soil recovery. There is always the possibility that he will be confronted with a prosecution for breach of the Preservation Order, and such prosecutions, even when not well-founded, are costly to defend.

Having stressed the seriousness of the problem, both from settlement damage from standing trees, and damage resulting from soil recovery when trees are felled, it is as well to recognise that there may be no need to prohibit development on well-wooded shrinkable clay soils provided suitable precautions are taken. For most developers the recommendations made in *BRE Digest 63* of allowing the site to regain water after trees have been felled is uneconomic because it would take too long. To avoid this delay the following alternative methods may be considered:

(a) Anchoring the building by using reinforced concrete bored piles sleeved from the ground over their top 3000 mm (10 ft) and providing suspended floors. Beams spanning between the piles must be well clear of the ground surface.

(b) Using flexible framed construction without brickwork or plastering.

(c) Making the building rigid by either constructing a basement or reinforcing the foundations and brickwork. Buildings erected close to standing trees should be supported on piles of sufficient depth, and for small structures bored piles are likely to be the most economical. Where nearby vegetation is insignificant, strip footings 1050 mm (3 ft 6 inches) deep are usually adequate.

On one particular site it would have been necessary to fell a large mature oak to permit development. There was a similar tree in the neighbour's garden, which the developer could not require the neighbour to fell until damage had actually been caused to the new development. It was found that short-bore piles, 2400 mm (8 ft) in depth, reinforced in tension, and with suspended ground-floor floors about 150 mm (6 inches) clear of the oversite, would be an entirely satisfactory solution. Moreover, provided the architect and piling specialist co-operated at the design stage, the cost of short-bore pile foundations, even for a single house, would be very little more expensive than traditional foundations 1050 mm (3 ft 6 inches) deep. *BRE Digest 298* stresses the unwillingness of developers to use pile and beam foundations in potential ground heave situations. The Digest points out that the deepened trench-fill, preferred by developers in this situation, is vulnerable as it is subject to lateral movement, rotation and large uplift forces. *Digest 298* suggests that for general design guidance for house foundations in swelling soils, reference should be made to *Digests 241* and *242*.

Recognition of tree root damage

Both professionals and householders are sometimes in ignorance of the influence of the roots of vegetation on foundations until a major settlement occurs. Then, confronted with damage, they tend to jump to the conclusion that trees are responsible, particularly if there is a near-by Lombardy poplar. This species, being fast-growing and retaining a compact shape has probably figured in more cases of settlement damage from tree root action than any other, because frequently it has been planted as a screen to hide development. The position, however, is not as simple as this, and not all fractures in brickwork are settlement damage.

In investigating these cases it is essential to establish that the soil is a shrinkable clay soil. Although a geological map should be consulted, the precise nature of the soil under the foundations must also be determined from excavations taken to the bottom of the foundations. This will establish whether the foundations are in firm shrinkable clay and not made-up ground or any other soil possessing inadequate load-bearing properties, for example, seams of soft yellow clay, such as occur in parts of Chingford, Essex. Excavations will also determine the depth of the foundations, which may well be criticised later if they are less than 900 mm (3 ft) in depth. Such criticism is not well-founded if they are 375–450 mm (15–18 inches) or more in depth, as this is below the depth where soil movements can be induced by climatic conditions alone.

Where the foundations are exposed it is essential to collect and identify any roots present under the foundations, remembering that it is not necessary to find large roots, although these are easier to identify. Settlement damage is caused by the extraction of moisture from the shrinkable clay in excess of what is recovered by way of precipitation of rain in abnormally dry seasons, and extraction of moisture is done by the fine root hairs. Sometimes while there may be, say, a poplar in the neighbour's garden, there are also trees or large shrubs in the garden of the damaged property as near as, or nearer to, the house than the neighbour's tree. Hence, quite extensive excavations are needed to ensure collecting representative samples of all roots. It is not always possible to link a particular item of damage with a particular root, and assessing damages in such cases has to be on a percentage basis, say 60 per cent to the client's vegetation and 40 per cent to the poplar, depending on the proportions of different roots found.

Excavating often produces surprising evidence that may greatly complicate the making of a claim. For example, if there are several trees of the same species, it is not possible to state categorically that a particular root has grown from a particular tree merely because the root appears to be growing in the line of that tree. Roots may, and often do, change direction. In an extreme case, roots were found to cross a boundary at a depth of about 1200 mm (4 ft) but, almost immediately, to change direction, growing parallel with the house, and diving to a depth of 3000 mm (10 ft). These roots were traced for a distance of about 10.5 metres (35 ft), parallel with the house, and then they changed direction, growing towards the house, but at a depth of 2100 mm (7 ft), and they were finally traced to within about 1500 mm (5 ft) of the house. In another case there was a row of poplars opposite a flank wall, and about 16.5 metres (55 ft) away. The position for the first excavation was selected close to a large fracture in the wall, expecting to find poplar roots. Instead, the majority of the roots proved to be oak from a tree in the garden of the damaged house, and 19.5 metres (65 ft) from the excavation.

Other complications that may arise are the finding of water in an excavation, when the defendant's expert, not unnaturally, endeavours to argue that the damage cannot be tree

root damage because there is obviously no shortage of water. The weather immediately preceding digging is important. The excavation will initially be in the back-filling of the trench
removed for the original foundations, and such ground forms a natural watercourse when
the water table is high following prolonged wet weather. This does not alter the fact that in
a prolonged dry spell, when the water table is low, the clay under the foundations will be
dried out by the roots in the clay. Water in the trench may be found from two other sources,
storm water drains that have been breached by roots and mains water leaks possibly as a
result of settlement. It is helpful to call in the Water Authority's Inspector, whose 'listening'
staff will detect leaks, or alternatively, the Inspector may be able to detect whether the water
is from his Board's supplies without the necessity for analysis. Fractured drains, or leaks
in the mains water supply, are often the explanation for only relatively slight damage to a
property in spite of the close proximity of trees. The explanation is that a constant source of
water may go a long way towards satisfying a tree's requirements. Several cases are known
where the re-laying of defective drains or repairing leaks in a mains water supply, without
felling near-by trees, have resulted in extensive damage in the next dry season because the
extraneous source of water had been cut off.

Excavations may also reveal the presence of underpinning, the existence of which
was unknown to the present owner and which may not have been documented. Having
established that the soil is a shrinkable clay, that there are roots growing up to and under
the foundations and that the foundations are below the depth where soil movement may be
caused by climatic factors alone, it remains to consider whether the damage is characteristic
of damage from tree root action and, finally, when the damage occurred. Typically, there
is damage not only to external walls but also to internal partitions and to ceilings, often
more extensive on the upper floors than on the ground floor. There are likely to be cracks
at junctions of walls and ceilings, gaps between the skirting and the floor that cannot be
explained by normal shrinkage of floor joists, and picture rails and skirtings in the angles
of rooms at different levels on adjacent walls. Corners of window and door openings, where
stresses are concentrated will also show cracks, see figures 14 and 15.

When the damage occurred is a vital question, because unless cracks or fractures first
appeared during or shortly after an abnormally dry season, they are unlikely to have been
caused by tree roots. The lady of the house is usually of more assistance than her husband
in remembering when damage first occurred. The dates of holidays are helpful, because
damage usually occurs late in the summer or early in the following autumn. Once severe
damage has occurred it is likely to be progressive, but further substantial damage usually
only occurs in a subsequent, abnormally dry year.

Claims for damages

It has been firmly established that trespass of roots from adjoining land that cause damage
to a neighbour's property gives rise to an action in nuisance, and substantial damages have
been awarded by the courts to the offended party. There is no certainty that a plaintiff
will succeed even if he can show that the soil is a shrinkable clay, that the foundations are
of adequate depth, that there are roots growing up to and under the foundations, that the
damage is characteristic of tree root damage, and that it occurred during or shortly after
an abnormally dry season.

Dispute as to the true position of boundaries may enable a defendant to repudiate ownership of a particular tree. This not infrequently arises in urban areas where a developer

has acquired land and built several houses, which are subsequently sold off separately, resulting in a strip of land being excluded from the deeds of all of the separate properties. If the offending tree is on this piece of land, it may be impossible to establish ownership. If several trees are involved, they may make for a multiplicity of defendants. Change in ownership of offending trees is another complication as it may be necessary to relate the items of damage to the particular times that different owners were in possession.

Erosion and movement of subsoil

Where trees are not present, leaking water mains or fractured drains on their own can cause severe damage to a property. In the case of a house in Chislehurst, see figure 14(iii), back-fill was washed away by leakage from the water mains. This action resulted in the creation of a large void under the party wall foundation, which eventually settled, causing cracks to appear in the wall above. The fact that the house was also built over part of a disused quarry exacerbated the failure. In addition to leaking services, sulphates within the subsoil or back-fill material can attack the concrete foundation, reducing its structural integrity and creating the conditions for settlement to occur. Houses built on sloping sites may also suffer movement where the subsoil is unstable or where the presence of ground water increases the likelihood of slippage. This is most likely during wet seasons when clay soils are present and where slopes are severe.

Mining subsidence

Houses built above mine tunnels may well suffer subsidence over long periods. Often buildings subside initially when tunnels are bored and then recover to a level below their original position after a period of time. Cracking patterns tend to be random in position as shown in figure 15(iii). A surveyor, working within a mining area, must be aware of the problem and the indications and consequences of mining subsidence on housing. A number of industrialised building systems for housing have been developed to overcome or reduce the effect of ground movement, caused by mining works. The surveyor should know of the various systems and be able to recognise each system when inspecting properties.

4 Floors

Floor construction is of paramount importance in regard to both the load-carrying capacity of floors, and in relation to the stability of the internal and external walls. Floor joists may provide essential ties for walls that might well not be adequate in themselves were they free-standing structures. For example, a 215 mm (one brick thick), free-standing boundary wall 6 metres (20 ft) high, would not be stable in itself (even assuming it had adequate foundations) whereas many houses at least 6 metres (20 ft) from ground level to eaves, with walls one brick thick, are still perfectly stable half a century later. This is because the cross-walls of a house, and the floor joists, coupled with the roof load, make for rigidity that does not exist with a free-standing wall.

The important consideration of a floor's load-bearing capacity will depend on several factors such as floor type and construction, span, length of bearing, age, condition and so on. One important factor is the distribution of loads upon the floor, loads from such things as baths, tanks and partitions. Partitions built off floors will also be of concern to the surveyor, for example, if work is necessary to repair a floor then the demolition and rebuilding of such partitions will add to the expense.

The surveying of floors calls for examining each floor in turn, and room by room, in a regular sequence. On the topmost floor, immediately under the roof voids or attic, there may be the additional task of checking that the trough and centre gutters, located in the course of investigating the rainwater disposal arrangements, have not given rise to outbreaks of fungal decay in the rooms underneath. When inspecting the ground and basement floors, the surveyor must also be on the lookout for evidence of rising damp and, if the ground-floor rooms are of appreciable size, evidence of any sagging of partitions in the rooms above.

Underfloor ventilation

In smaller properties with suspended timber floors on the ground floor, and no basements or cellars underneath, it is important to check that through ventilation and cross-ventilation

32

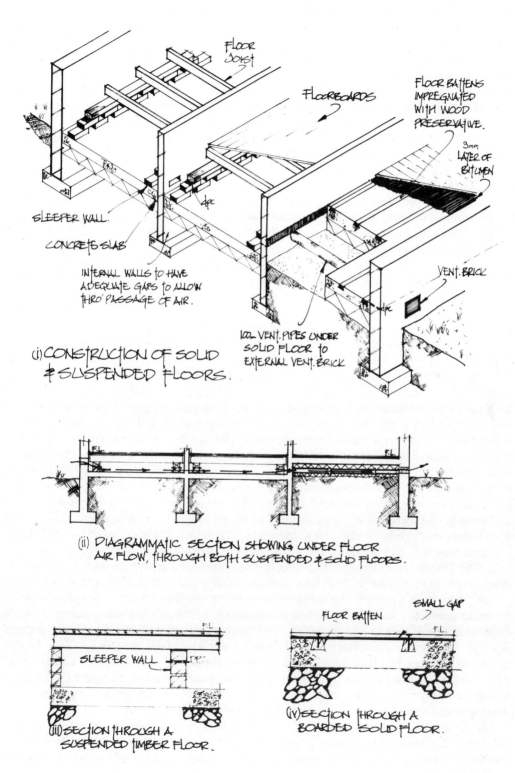

FLOOR JOIST

FLOORBOARDS

FLOOR BATTENS IMPREGNATED WITH WOOD PRESERVATIVE.

3mm LAYER OF BITUMEN

SLEEPER WALL.

CONCRETE SLAB

INTERNAL WALLS TO HAVE ADEQUATE GAPS TO ALLOW THRO' PASSAGE OF AIR.

VENT. BRICK

100. VENT. PIPES UNDER SOLID FLOOR TO EXTERNAL VENT. BRICK

(i) CONSTRUCTION OF SOLID & SUSPENDED FLOORS.

(ii) DIAGRAMMATIC SECTION SHOWING UNDER FLOOR AIR FLOW, THROUGH BOTH SUSPENDED & SOLID FLOORS.

SLEEPER WALL

(iii) SECTION THROUGH A SUSPENDED TIMBER FLOOR.

FLOOR BATTEN

SMALL GAP

(iv) SECTION THROUGH A BOARDED SOLID FLOOR.

Figure 6 Solid and suspended floors

under the floor are adequate and that they are not impeded by areas of solid floor and internal partitions taken down to the oversite in solid brickwork. Obviously, in semi-detached houses there will be no cross-ventilation, and the adequacy of ventilation from front to back is all the more important. Checking that air bricks are not obstructed does not, of course, ensure that there is through ventilation under the floor between rooms, as there may be no voids in the partitions that have been taken down to the oversite. Where practicable, floorboards to be taken up should be those against internal partitions and in the recesses between projecting chimney breasts. Figure 6 ((i)–(iii)) reproduced from an early BRE Digest, illustrates the good practice that was advised in the 1950s for suspended timber floors with adjoining solid floors. It also indicates the points to look for when inspecting suspended floors, including the provision of ventilation by means of pipe ducts through the adjoining solid floors.

Figure 6(iv) reproduced from the same Digest shows a typical detail of a solid floor but the surveyor is unlikely to be able to check such details in the course of a reconnaissance survey. The floor finish on solid ground floors most likely to give trouble is the boarded or strip floor on battens. In the older properties the battens will almost certainly not be pressure-treated, and there is unlikely to be an impervious membrane in any position under the boarded floor. There is frequently a small gap between the top of the oversite concrete and the underside of the floorboards, see figure 6(iv), and, with no impervious membrane, such floors have a very uncertain life, particularly if the boards are a non-durable species such as European redwood. High relative humidities can build up under the boards, creating ideal conditions for the development of dry rot. Such solid boarded floors, if of softwood, require very careful scrutiny, including tapping and probing for any signs of decay.

Unless the floors are hardwood strip floors, laid by a specialist flooring contractor, it is most unlikely that the flooring will have been kiln-dried prior to laying, and the width of gaps between the boards or strips should be noted. If the boards are still tight in properties or new extensions more than a few years old, the surveyor should suspect that either the under-floor ventilation is not adequate or that the builder has made a great error and that there is no oversite concrete. Boards or strips are sometimes so tight that the floor can be seen to be wavy, but curvature in the boards is sometimes only apparent if one rubs the palm of one's hand over the floorboards. If floors are of some age and the floorboards are still tight, there is a considerable risk of there being fungal decay in the joists or wall plates underneath, and therefore it is imperative to investigate further. Gaps between skirtings and floorboards may provide a clue, but a search should also be made for areas of patched floor-ing, particularly in bay windows and in the recesses between projecting chimney breasts.

In older houses, short lengths of skirting should arouse suspicions because at the time such houses were built, timber in good lengths was usually readily obtainable, and the existence of short lengths is a pointer that it has been necessary to effect repairs. The surveyor must be on his guard against being misled by the piecing in of the skirting where partitions have been taken down, or wall fittings have been removed. It is not only in old houses that there is a risk of fungal decay developing in the joists and plates under the ground floor. Severe decay has been known to develop within eighteen months of completion of construction when the first occupants cover floors with impervious floor coverings and where low quality, unseasoned and non-treated timber has been used. The risk is not great in modern estate development because ground floors are almost invariably solid floors.

The floors of kitchens and cloakrooms should receive particular attention as not only do many solid ground-floors impede through or cross-ventilation of the floor joists, but there is the possibility that quarry tiles have been laid on a screed on top of a timber sub-floor. In addition, the covering of kitchen floors with linoleum or similar impervious floor coverings, if this is taken up to the wall under a sink unit, may also give rise to a higher risk of decay. The floor covering should always be turned back under the sink to permit inspection of the flooring.

It is often impossible to inspect the floor or wall at the back of the sink unit in the course of a reconnaissance survey because of fitted cupboards under the sink with a plywood/chipboard base and back to the units. If this is the case the fact must be mentioned in the survey report, stressing that it has not been possible to inspect the most vulnerable part of the floor. As already stated, modernisation of kitchens by laying thermoplastic coverings on the original timber floor often increases the risk of fungal decay. Even with tongued and grooved flooring there is a prospect of some air movement by convection, through the floor, but, if the top surface is effectively sealed by some form of impervious covering stuck down to the existing timber floor, such air movement can no longer occur. In such circumstances the relative humidity of the air under the floor is liable to rise to critical levels, and the moisture content of previously 'dry' timber will also rise until it reaches levels that can support the growth of wood-rotting fungi. Moreover, depending on aspect, the nature of the subsoil, and the amount of heating enjoyed in the kitchen, condensation under the floor may occur from time to time leading to saturation of the flooring joists. These conditions can cause the development of fungal decay in as short a period as eighteen months. Hence the necessity for warning a client of the decay hazard created by sticking impervious floor tiles or floor coverings stuck down to existing timber floors. The sealing of floors with impervious floor coverings on the ground floor is not confined to kitchens. To provide smooth floors to receive carpets, sheets of hardboard are often laid over the old boards, and this material is equally effective in preventing air movement through the gaps in the floorboards.

It is important to check that all air vents are of the same pattern, and if the house is one of several similar houses in the same road, that the number and pattern of the air vents in the property being inspected are identical to those in the neighbouring properties. In the course of the external inspection, the surveyor should have noted where new, larger, or additional air vents have been inserted and the floors and skirtings in the vicinity of such vents should be carefully investigated. It is more than likely that new floorboards, or new lengths of skirting, will be found opposite the new vents, and floorboards should be taken up to ascertain whether any joists or wall plates have been renewed, and that there is no evidence of recurrence of any trouble. Unfortunately, even if previous attacks under floor in one part of the room have been dealt with adequately, this is no guarantee that attacks do not exist at the opposite end of the room, from the same causes—for example, ingress of water at other points, or poor underfloor ventilation. This is particularly likely with true dry rot *(Serpula)* infection.

If fruiting bodies have been produced at one end of the room, spores will drift across the room, and if they alight on timbers whose moisture content has been built up to 24–28 per cent, there is a considerable risk of several separate outbreaks becoming established. It is this possibility that detracts from the value of most guarantees given by commercial dry rot treatment firms, which are often transferable to a new owner. As a general rule these twenty-year guarantees are limited to the actual timbers treated when the guarantee was issued, and timbers in the same room that were not treated are not covered by the guarantee. This should be pointed out to the client if the vendor produces such a guarantee as evidence that a previous outbreak was dealt with by a responsible firm. The surveyor must not himself rely on the guarantee, for he must satisfy himself that there are no other outbreaks in the same, or adjacent, rooms. The initial attack may be close to the external front wall, or against an exposed flank wall, or in the vicinity of a faulty downpipe attached to the wall. Where this is the case, floorboards should be taken up in the area where repairs have been carried out, and against the partition at the other end of the room, and in the recesses on both sides of the chimney breast.

Beetle infestation

Apart from looking for infestation in floors, skirtings and picture rails, as on the upper floors, there are two points on the ground floor that appear most vulnerable—these are cupboards under stairs (including the backboards for meters and fuse boxes) and larder shelving. If the soffit to the staircase is not plastered, the glued blocks in the angles of treads and risers appear to be particularly vulnerable to attack. It is always worth while to clear everything out of the cupboard under the stairs to permit a thorough inspection. In the larder it will be possible to see from the underside of the shelving whether there is sufficient sapwood in it to warrant removing everything to permit examination of the shelf tops.

Fill

Since the Second World War, increased demand for housing has encouraged developers to use land which in the past was thought unsuitable for building. Examples of such marginal sites were those that sloped severely, supported large numbers of trees, had drainage problems or required large amounts of fill.

   With respect to the last difficult site, one common problem of post-war housing with solid floors was that of settlement, this being mainly due to the inadequate consolidation of the sub-floor filling and/or the use of unsatisfactory fill materials. The surveyor when inspecting such properties should search for evidence of 'dishing' or 'sloping' of the floor and/or the existence of gaps between the skirting board and the floor. The surveyor should be aware that gaps may not always be found where settlement has occurred, for it is not unknown for skirting to be removed and replaced at a lower height to cover up the defect. By using a one metre spirit level the surveyor should be able to recognise any deficiency and report this to the client.

Upper floors

In a reconnaissance survey an important consideration is the soundness of the floor whereas in a detailed structural survey determining its load-bearing capacity is likely to be the primary object. As regards soundness, the important considerations are the extent and significance of any beetle infestation and/or fungal infection in the timber members.

   The first matter to be determined is whether the joists are parallel with, or at right-angles to, the external walls. In smaller properties it is usually relatively simple to establish this from the direction of the floorboards because there is unlikely to be a sub-floor. If however, the floorboards are tongued and grooved, the surveyor is in difficulties because he will not be able to take up any flooring without breaking out at least one floorboard. If joists are parallel to the external wall, the risk of decay below the floor is governed largely by whether or not the first joist in is tight to the wall or 25 mm (1 inch) or so back from the face of the wall. The surveyor should be able to determine this from the position of the nails, provided the boards are face-nailed. If the flooring is tongued and grooved and secret-nailed, the surveyor's only hope is to probe with a piece of wire under the skirting in an effort to locate the position of the first joist in. With joists housed in the external wall, decay is most likely to exist where the ends of joists are housed in exposed solid walls or in walls close to external downpipes (when the condition of the face of the wall behind the downpipes is all important). If the surveyor is not able to take up floorboards he should examine the skirting very thoroughly, feeling for any curvature in the skirting, and probing it where it is close to such damp walls. Jumping on the floor is not a reliable test because the amount of give in a floor will depend on many factors such as the relationship of the depth of the joists to their span, the weight of the surveyor, where the floor is tested and so on. Only with experience can a surveyor use this test with confidence, and in those cases where a problem may exist, further scientific tests should be carried out.

Special attention has been drawn in respective chapters to the hazards presented by balconies, flat roofs over bay windows and other projections in the vicinity of the floors being inspected. If the surveyor is not in a position to lift floorboards adjacent to such features in the course of his reconnaissance survey, he should examine balcony floors and areas of flat roof very thoroughly, noting the extent of repairs, and whether the fall is away from or towards the external walls. If there are 'French windows', the thresholds should be probed, as should the sills of any windows close to balcony floors or areas of flat roof. If the thresholds or sills are oak, and the timber is soft, the likelihood of there being decay in the floor timbers should certainly be suspected. This is because the joists may well be of softwood, which is more prone to decay than oak sills. The surveyor must make it clear in his report that he was concerned about a possible defect and the extent to which he was able to investigate. He should endeavour to give some indication of what sum for repair might be involved, if his suspicions that decay is present prove to be well-founded. The prospective purchaser can then decide whether he is prepared to gamble, or whether he is willing (if the vendor agrees) to incur the cost of opening up and making good any damage done.

Framed floors

In a detailed survey, opening up against external walls is essential to establish the condition and dimensions of joists and wall plates. In the larger properties it is to be expected that the floors are framed floors—that is, the floor construction consists of one or more large beams spanning the shorter dimension of the room, with the floor joists at right-angles to the beams, and housed in them, see figure 7. The main beams are usually of timber or steel, but occasionally are of cast or wrought iron. In buildings 100 years of age or more, the construction of framed floors is often very complex, with a combination of single beams of timber or cast or wrought iron. In such older buildings, particularly if spans are large, various types of composite beams may be found (that is, two beams bolted together, with a distance piece between, or flitch beams) as shown in figure 8 ((i)–(iv)). The framed floors of large rooms are likely to consist of two or more main beams, in which the principal joists are housed at centres varying from 900 to 1200 mm (3 to 4 ft), up to 2400 mm (8 ft) or more, with secondary joists at 300–350 mm (12–14 inch) centres, at right-angles to the principal joists, and notched over them. In older buildings, neither the beams, principal joists nor secondary joists, even in one room, are likely to be of uniform sectional area.

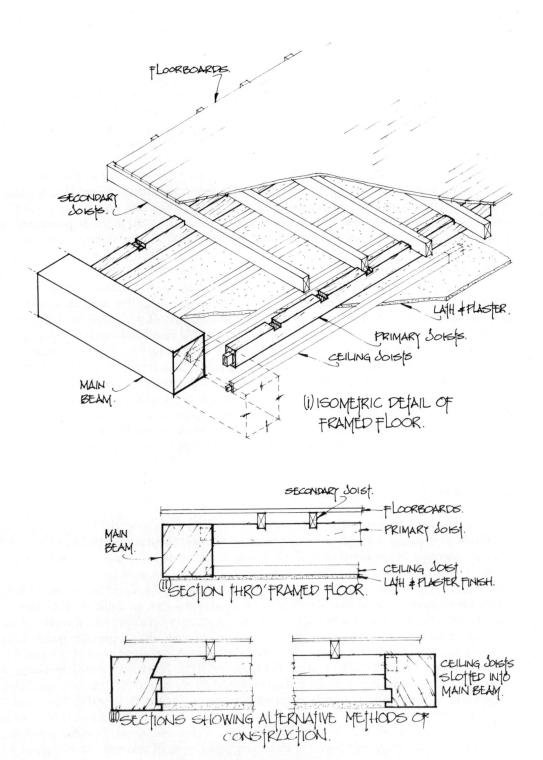

FLOORBOARDS.

SECONDARY
JOISTS.

MAIN
BEAM.

LATH & PLASTER.

PRIMARY JOISTS.

CEILING JOISTS

(i) ISOMETRIC DETAIL OF
FRAMED FLOOR.

SECONDARY JOIST.

MAIN
BEAM.

FLOORBOARDS.

PRIMARY JOIST.

CEILING JOIST.
LATH & PLASTER FINISH.

(ii) SECTION THRO' FRAMED FLOOR.

CEILING JOISTS
SLOTTED INTO
MAIN BEAM.

(iii) SECTIONS SHOWING ALTERNATIVE METHODS OF
CONSTRUCTION.

Figure 7 Framed floors

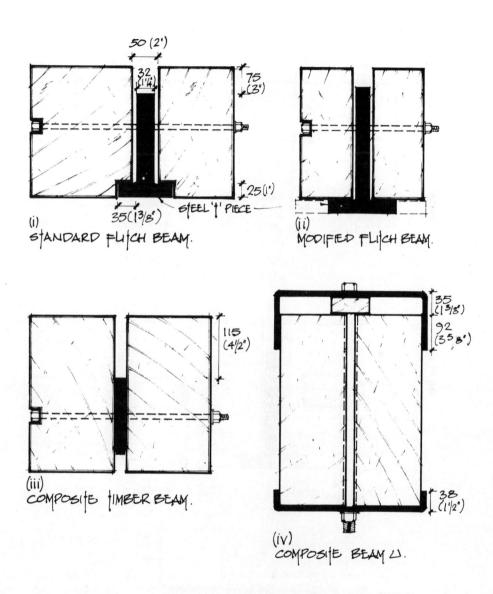

Figure 8 Flitch beams

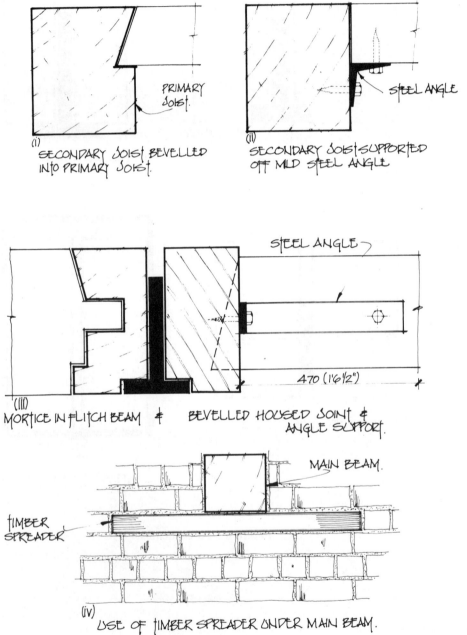

Figure 9 Timber housings and supports

The nature of the housing for the principal joists in the main beam is also likely to vary. In addition to tenons of conventional pattern, others of unusual shape and size may be encountered, or the principal joists may be simply slotted into the main beam as shown in figure 7((ii) and (iii)) and figure 9((i)–(iii)). The simple slotting of principal joists into the main beam has certain advantages over conventional tenons, in that shrinkage of the main beam does not result in the principal joists becoming dependent only on their tenons, probably no more than one-third of the depth of the joists. Many of the elaborate housings involve huge mortices in the main beam, and, as a result of shrinkage, the fit of one member in the other is no longer tight, reducing the effective cross-sectional area of the supporting beams appreciably.

In older buildings additional services are likely to have been installed from time to time, and plumbers and electricians are given to notching beams with a total disregard for the weakening effect of the notches. As the original main beams in older properties were nearly always of inadequate size, loss of effective sectional area from mortices and notches, coupled with the destruction of sapwood edges by old beetle infestation, can be extremely critical, particularly if deep notches occur in the middle third of the span. It is not unusual to find that the main beam has been still further weakened by a groove cut close to the bottom or top edge of the beam to house the ceiling joists after the main floor structure has been framed up, see figure 7(iii).

To permit mathematical analysis of the floor construction, it is necessary in a detailed structural survey to plot all the members accurately on a measured drawing, and, as the construction is often complex, this frequently necessitates taking up the whole of the flooring, and even patches of ceiling under, to obtain the necessary data. If the floor construction is uniform, that is, consisting of a series of beams at 2750–4600 mm (9–15 ft) centres, with subsidiary joists at right-angles to the main beams, it is still necessary to take up the floorboards both sides of the beams to determine the quality and condition of the beams, and whether the subsidiary joists are still properly housed into the main beams.

In the course of detailed structural survey it is essential to check the condition of the ends of the main beams housed in external walls. These initially may have had a spreader in the wall under the beam figure, see figure 9(iv), but it is not unusual to find that the spreader has been cut out because of decay, and the void left has been built up in brickwork. In consequence, the beam imposes a point load on the wall, and this can lead to the development of fractures in the wall under the beam. Repairs carried out in the nineteenth century were often indescribably shoddy.

Finally, the extent of beetle infestation must be investigated. In oak timbers the important borer is, of course, the death watch beetle, because this infestation is almost always secondary to fungal decay. If the surveyor learns that the timbers have been treated by a commercial servicing firm he should ascertain precisely what work was done. *In-situ* insecticidal treatments carried out by inexperienced operatives can result in important structural members being more severely weakened by the remedial measures than any damage that might have been done by the beetle infestation. For example, in one case all the main floor beams of a school assembly hall had been bored close to both edges, and almost for the full depth of the beams, at regular intervals throughout their length; the borings were 8 mm ($^5/_{16}$ inch) in diameter. Since the beams were oak the boring would not have materially assisted the spread of the insecticide in the timber, but the effective width of the beams had been reduced by about 50 mm (2 inches) which had an important influence on the load-carrying capacity of the floor.

Bath and WC floors

Fungal decay in bathroom floors from plumbing leaks and persistent condensation should always be suspected, particularly where floors have an impervious covering, or worse, where impervious tiles are stuck to the timber floor. Wet rot and furniture beetle attack are often found around the base of the WC pan.

The surveyor is in a real difficulty when he encounters a solid bathroom floor in a house where the floors of other rooms on that floor are ordinary timber floors. In a reconnaissance survey, there is usually no means of determining how the floor has been constructed. The likelihood that it has been properly constructed, with an adequate damp-proof membrane, cannot be assumed. As often as not in smaller properties, the structural floor members will be timber joists with mass concrete between the joists. This then supports a screed of up to 38 mm (1½ inches) thick over the whole floor, on which the finish of tiles will have been laid. Water is very likely to have found its way through the tiles and screed, so that the timber joists may have become infected with dry rot. All the surveyor can do is to search for any symptoms of dry rot anywhere in the bathroom or in the ceiling underneath, and for cracks in the tiles. If there is a panelled bath the bath panel should be removed so that the state of the floor under the bath can be inspected. In his report, the surveyor should draw attention to the advisability of establishing the construction of the floor, if necessary, by breaking up a small area under the bath. The risk is that by the time timber dry rot becomes manifest, the floor may be in imminent danger of collapse.

Floor repairs

In old houses with wide oak and elm floorboards it has often been necessary to patch such floors because of excessive sapwood edges, which were destroyed by *Lyctus* and furniture beetle infestation over the years. This is no more than normal wear and tear. The patching of softwood floors of later date may be of much greater importance, depending on where the patches are located. For example, two or three 'new' floorboards against external walls and particularly under windows, may be an indication that it has been necessary to deal with an outbreak of either dry or wet rot. It is imperative to lift such new boards and to make a thorough examination of the joists and wall plates underneath. This precaution has often revealed that the repair work has not been carried out properly, and it may lead to the discovery of a continuing attack that the surveyor could not possibly have detected in the course of a reconnaissance survey, at any other point in the property. One such case led to the discovery that dry rot existed in the ground and basement floors, but only after linings around window openings were dismantled and substantial areas of wall plaster had been hacked off. The remedial works at that time cost close on £2000. It is just such features, apparently insignificant in themselves, that should put a surveyor 'on enquiry'. The inclusion of 'escape' clauses in the survey report would be unlikely to provide any defence if the areas of patched flooring ought to have been seen. Admittedly vendors are often not prepared to allow fitted carpets to be lifted, but if there are factors that give rise to suspicions, for example, scoured brickwork behind downpipes, or hardboard or plywood panels under windows, then the surveyor should advise his client not to complete negotiations for purchase if permission is not granted for a reasonable amount of opening up, unless the client is prepared to risk having to spend substantial sums of money later.

Attic floors and conversions

Since attics and store rooms are accommodated within the roof, it is important to check that their formation has not weakened the roof construction. To obtain adequate head room in the attic, collars almost invariably have to be fixed more than halfway up the roof slope. In these cases, resistance to the outward thrust of the roof is dependent on the floor or ceiling joists tying in the feet of the rafters. Provided these joists are at right-angles to the roof slopes, and that they are fixed to the feet of the rafters, spreading of the roof will be restrained. The joists must be adequate to support the loads from the internal partitions parallel with the joists, and the weight of the ashlaring walls. Further, the rafters must be of adequate section, and adequately supported by purlins, to ensure that the roof load is not transmitted via the ashlaring walls to the floor. The surveyor must also check that any trimming of rafters, to accommodate dormer openings, has not weakened the roof construction. In smaller properties with simple double pitch roofs, there should be no problem, but in old properties it may be found that the floor is framed—that is, there is a main longitudinal beam in which subsidiary joists at right-angles are housed. As with such framed floors at lower levels, the main beam, although often of considerable size, may not be adequate for the very substantial floor load it has to carry. Moreover, the main beam is not always parallel with the ridge, but at right-angles to it. In this case the feet of the rafters may not be tied adequately by the subsidiary joists.

In large houses the attics frequently present more complex structural problems. Often the roof slopes are mansard slopes, with a large area of flat roof between, supported in part off load-bearing internal partitions. In houses 100 years old or more, the floors of large rooms will almost certainly be framed, where the main beams will often bear on to the external walls, usually under a parapet gutter. It is imperative to check the condition of the ends of these beams, and the condition of the wall plates and the feet of rafters, even in the course of a reconnaissance survey. In a detailed structural survey the position and dimensions of all the floor timbers must be plotted on the floor plan of the attic floor. The housing of the subsidiary joists in the main beam must always be investigated. Moreover, the main beams of ancient buildings were never seasoned timber when fixed, and a 300 x 300 mm (12 x 12 inch) beam, particularly of oak, will have shrunk appreciably with the passage of time, and serious splits or shakes may have developed. It is not unusual to find that as a result of shrinkage, the subsidiary joists are no longer adequately housed in the main beam.

When the whole of the roof void has been used for attic accommodation, one room is likely to house the cold storage tank or tanks. It should be borne in mind that plumbing on the upper floors of older properties may well have been installed long after the house was built, and hence the adequacy of the tank room floor for supporting the substantial load of a large tank warrants close investigation. In the larger country houses, particularly if the water supply originally was not mains water, the cold storage tanks are often of alarming capacity. Tanks of 9000–13 600 litres (2000–3000 gallons) capacity are by no means exceptional, representing a dead load of 9–13.6 tonnes, apart from the weight of the tank. Such large tanks may be of lead or even slabs of slate, adding still more to the load on floors not designed for such abnormal loading. It will usually be found that, at best, an attempt has been made to distribute the weight of the tanks, although often by means that will horrify a structural engineer. Timber in the vicinity of the tanks must be examined for evidence of decay, and the route of overflow pipes to the outside should be followed. This precaution on one occasion brought to light the fact that the end of the main beam of the framed tank room floor was in an advanced state of decay, as indicated by the size of the fruiting bodies of *Serpula*.

The surveyor will obviously look for evidence of furniture beetle infestation in the attic floors. If the property is more than fifty years of age it is more than likely that there will be no continuing active attack on articles in the attic rooms. In light rooms, as opposed to roof voids and the voids behind ashlaring walls, the colour of the frass is a useful indication as to whether the infestation is still active, and a search for dead beetles may be rewarding.

Where the attic floor covers a large area there are likely to be lantern lights, and these require close inspection, including a further visit to the flat roof above to check on the condition of the flashings to such lights. It will often be found today that attic floors of many of the larger country houses have not been occupied for some years, being used only for storing discarded furniture, packing cases and other articles that may have been much better burned. Since no one has occasion to visit the attic floor regularly, serious structural deterioration may occur without the owners being aware of anything amiss, and this is something that the surveyor should have very much in mind when inspecting unoccupied attic accommodation. Often such floors were used for staff rooms and elaborate precautions may have been taken to render them sound-proof, usually by pugging supported on battens or thin boards 50–75 mm (2–3 inches) below the floorboards. This pugging is normally hair-lime mortar, but many different materials have been used, including sawdust. When there is absorbent pugging under the attic floors, rainwater may often have been getting through defects in the roof covering at several points over a long period, without the owners being aware of the fact. Hence, unless the survey is being carried out on a wet day, the flooring of all unoccupied attic rooms should be searched for tell-tale water stains, and such stained floorboards must be taken up and a thorough search made for any traces of fungal attack. The points of ingress of water must also be traced.

In one case sound-proofing took the form of a thick layer of felt under the attic floors which, with the passage of time, had become the breeding ground of clothes moths and carpet beetles in enormous numbers. Insecticidal sprays in the bedrooms underneath having failed to keep the moth population in check, entomologists of the Pest Investigation Laboratory were consulted. They recommended inspecting the voids behind ashlaring on the attic floor, where the felt introduced for sound-proofing was found to have been reduced to little more than particles of fluff. It was necessary to take up the whole of the flooring to remove the remains, sprinkling insecticidal powder in large quantities between the floor joists before re-laying the flooring.

On completing his survey of the attics, the surveyor should check that all windows, and doors or dormers giving access on the roof, have been shut and properly fastened.

Skirtings

The importance of checking on the condition of skirtings against external walls must be stressed, because buckling is often the first indication of the presence of fungal decay. In close-carpeted rooms the skirtings may also provide evidence of the existence of furniture beetle infestation. As a paint finish provides some protection against egg-laying of wood borers, any unpainted skirting should be inspected for signs of attack. Skirtings should also be tested to ascertain whether they are of timber or plaster. If made of plaster when the floors are joist floors, this may indicate that old timber skirtings have had to be renewed in plaster because of earlier decay. The life of the timber floor in these cases should be suspect, even if no decay can be detected.

Cellars and basements

These must be inspected thoroughly, particularly if soil below ground level has not been dug out to provide an external open 'area' around the perimeter walls. In practice it is rare to find a continuous open area, and damp conditions are therefore common. This is because it is exceptional in older properties for there to be a vertical damp-proof course between the soil and the brick or masonry wall below ground level. If conditions generally are damp, the base of walls should be inspected in a strong light for any tell-tale marks indicating periodical flooding.

It is a costly matter to make a really damp cellar dry, because the only certain solution is to tank the cellar, usually with asphalt. This will entail raking out the joints in the brickwork or masonry to provide a key for a rendering coat, trowelled smooth, to receive the asphalt. It is then necessary to apply a loading coat, to keep the asphalt in position. The thickness of this depends on the degree of dampness and the depth of the cellar. Before tanking the walls the cellar floor must be dealt with similarly. That is, the floor must be prepared to receive asphalt, which is taken up the walls to ensure a continuous impervious membrane. The asphalted floor will require a loading coat of reinforced concrete of adequate thickness to resist the pressure of water when the water table in the soil is high.

If cellars are no more than about 2400 mm (8 ft) below ground level, and the water table is generally below floor level in the cellar, a 102 mm (4 inch) brick skin built about 50 mm (2 inches) back from the cellar walls, and back filled with concrete, is likely to suffice. If however, there is any likelihood of the water occasionally rising 2–3 metres (6–10 ft), the internal support for the asphalt needs to be a reinforced concrete box, designed to withstand the full thrust on the floor and walls. Some cellars are equipped with self-priming pumps to deal with recurring water percolation. Apart from the fact that these can break down, allowing water to rise dangerously high in the cellar, the cutting in and out of the pumps can be annoying if they are located beneath one of the reception rooms.

It is unwise to tank such a cellar in the manner described above unless the source of percolating water is first traced. If percolation occurs because of a rise in the water table following heavy rain, the effect of tanking the cellar may be to cause water to rise above the oversite in those parts of the house where there are no cellars, and the result of tanking is then to make the rest of the house damper than before.

If it is necessary to have a pump in the cellar, it is essential to make exhaustive investigations regarding the possibility of flooding. The past history of a property can be misleading if land that previously belonged to the house is sold off. If the land includes a stream, a previous owner has often been in a position to divert water by means of sluices, to prevent flooding of the house, but when these head waters are in separate ownership the new owner may be deprived of such facilities, and the risk of flooding becomes very much greater. Although it is helpful to make enquiries from local residents, the surveyor should not rely on such information but should consult the Water Authority for the area.

If there is no ceiling in the cellar the ends of joists housed in external walls should be probed. Frames to window openings and chutes should also be examined carefully, particularly for evidence of fungal decay. Some cellars have brick or stone vaulted ceilings which, structurally, may be entirely satisfactory, but this usually means that there is very little space between the top of the vaults and the underside of the ground-floor flooring. In such circumstances the surveyor should pay particular attention to the flooring above the vaulted cellar ceilings. This will be necessary to ensure that lack of ventilation under the floor is not causing the build-up of high relative humidities in the timbers. If the boards or strips are tight and the surface of the floor is at all wavy, there may well be fungal decay in the joists.

Cellars and basement floors

The problems likely to be encountered in basements are similar to those that occur on the ground floor of properties without basements, but rising damp, and decay in timber floors, skirtings, door linings, and frames are likely to be more general, even in otherwise well-built properties. Basements are invariably partly below ground level, but if intended to be used as living accommodation there is usually an open area all around the perimeter walls, below ground level. Frequently, however, these areas are not sufficiently deep to permit adequate underfloor ventilation for suspended timber floors and thus may encourage fungal attack. Such conditions may also cause secondary infestation by wood-boring weevil. The measures to be taken to overcome the decay will eliminate weevil infestation without the necessity for insecticidal treatment.

If there are no open 'areas', rising damp problems will extend much higher up the walls than in ground-floor rooms with no damp-proof courses, and large areas of wall plaster may well be perished. The surveyor should be suspicious of basement walls covered in whole or in part with 'matchboarding', a not infrequent solution fifty or so years ago when timber was relatively cheap. Although such boarding may be fixed to horizontal grounds, it is probable that the grounds, if not the matchboarding itself, will contain wet rot. The boarding should be probed for decay and even if none is located, the surveyor should nevertheless record that covering damp walls with matchboarding is not a permanent solution and the risk of decay developing later is great.

If it is intended to use basements below ground level as habitable rooms, the minimum satisfactory solution is to hack off existing wall plaster, applying 'Romanite' lathing or similar to the walls before re-plastering. The client should be warned that re-plastering in the basement, or for that matter, the laying of solid floors, will cause appreciable delay, because basement rooms are not as well ventilated as ground-floor rooms, and the drying of plaster, concrete, or even a new screed on an existing concrete floor, may take twice as long.

Evidence of patching of joist floors should be searched for and floorboards should be lifted to check whether they are original. If the boards are more than about 140 mm ($5^1/_2$ inches) on face, and the quality of the timber is good, the flooring is almost certainly of some age, but the type of nails should settle any doubt. Taking up floorboards also permits the depth between the underside of the floor and the top of the oversite to be determined, and the type and nature of the oversite. If the present floor is sound but not original, the surveyor should stress in his report that the further life of the floor must be regarded as uncertain and, ultimately, it will be necessary to renew the floor again, and then a solid floor should be substituted. An important factor affecting the life of boarded basement floor (apart from the type of floor covering) is the nature of the oversite. Original basement flooring has been found in houses eighty years of age where there was no concrete, and poor underfloor ventilation, but the subsoil was gravel. Similarly constructed floors in clay areas might well last not more than twenty years. The depth between the underside of the floor and the oversite affects the cost of laying solid floors, but the type of floor finish selected is the major cost factor.

Even if there is good oversite concrete but no membrane, and reasonable ventilation under the basement floors, the surveyor should warn against laying thermoplastic coverings on top of an existing timber basement floor. If there is no oversite concrete and poor underfloor ventilation, he should make it clear that a new owner would be well advised to budget for laying solid floors within a few years. In modernising an existing basement kitchen by installing elaborate wall fittings and a sink unit, the prudent course is to lay a solid floor at the same time.

In recommending the laying of solid floors, attention should be drawn to the importance of not bridging the horizontal damp-proof course in the walls. There should be an impervious membrane in the solid floor, either 500 gauge polythene on top of the hardcore, or a membrane sandwiched between two 75 mm (3 inch) layers of concrete, or one laid on top of the concrete but under the porous screed. Before laying polythene on hardcore, the latter should be levelled with sand or lean-mix to ensure that the membrane is not punctured when laying the concrete. If the membrane will be level with the existing horizontal damp-proof course in the walls, the joint should be raked out so that the membrane in the floor can be tucked into this joint to provide a continuous impervious barrier in both floor and walls. If, however, it must be positioned either below or above the damp-proof course in the walls because of the finished floor level required, the floor membrane must extend up or down the wall so that it can be tucked into that joint. These points are discussed in *BRE Digest 54* 'Damp-proofing solid floors'. Even if certain parts of the basement have apparently adequately ventilated boarded floors, there will probably be some areas of solid floor with a variety of finishes, for example, brick on edge, stone slabs, quarry tiles, or only concrete. Such areas are likely to interfere with through and cross-ventilation, and the surveyor should look in cupboards against internal partitions for grilles let into the floor. Grilles against internal partitions opposite external walls help to prolong the life of a poorly ventilated floor but, if located in cupboards, they may be of little assistance.

In the larger, older urban houses, areas of wood-block flooring may be encountered in the basement. The whole of such floors should be carefully inspected for loose blocks, traces of decay, and water stains. The adhesives used for sticking down the wood blocks tend to become brittle with age and if the oversite is damp, the life of a wood-block floor may be short. When it is necessary to take up and renew such a floor, cleaning off the remains of the adhesive, which is essential, adds appreciably to the cost.

Having checked on walls, floors, skirtings and external joinery, door frames and linings should be thoroughly inspected for any traces of decay. This will often spread up behind door linings, much higher than any rising damp, even up to the head of the opening because humidity levels will be higher in the voids between the back of the linings and the reveals of the door opening.

The services should be inspected as on other floors, but in addition, the surveyor must be on the look-out for drains under basement floors. They should be in cast iron or clayware surrounded in concrete. Manholes within basements should be of a superior type and be properly sealed. Often in older properties the drains under the floor are ordinary salt glazed clay pipes and the manholes may well not meet the requirements of the Environmental Health or Building Control Officers.

Underground room regulations

Basement rooms used in the past for human habitation may fall short of the requirements of present-day regulations, and the surveyor should point out that the Environmental Health Officer has powers to prohibit the use of some or all parts of the basement for human habitation unless substantial and expensive alterations are made.

Dampness may be stopped by laying solid floors, incorporating an impervious membrane, and replastering both external walls and internal partitions on 'Romanite' lathing or similar. Light and ventilation may cause difficulty if it is not possible to enlarge open 'areas' in front of external walls because such walls are too close to the street or the boundaries of the property. Local Authorities may require the execution of specified works or repair, or, if they are satisfied that the accommodation is unfit for human habitation and is not capable at a reasonable expense of being rendered so fit, may invite the owner to put forward his proposals for bringing the rooms up to the required standard. In certain circumstances the Local Authority may not be satisfied with re-plastering on 'Romanite' lathing, and may insist, in addition, on the provision of a horizontal damp-proof course. Until the work has been done, and approved by the Authority's officers, the basement accommodation cannot be used. Failure to satisfy the Authority may result in the issue of a Closing Order on certain rooms, or even the whole basement. Owners have often embarked on conversion operations without obtaining planning permission or building regulation approval, and these conversions may not be acceptable to the Authority's officers. A new owner intending to continue to use the basement as converted must be warned about this. The Authority's officers have powers to inspect even when it has not been necessary to call them in about proposed alterations. If an officer is refused permission to inspect, a formal application for permission may be served, and, in the last resort, the Authority can obtain a court order.

Sound-proofing of floors and partitions

The surveyor is not infrequently asked to advise on the sound-proofing of floors and partitions, for which there is rarely an easy solution and, in many properties, it is impracticable other than at prohibitive cost. As regards floors, there are two sources of noise to be circumvented, namely, transmitted noise, from walking overhead, and air-borne noises. The former can usually be reduced to an acceptable level with thick underfelts and thick carpets on the upper floors. The cutting out of air-borne noises is much more difficult and is dependent on mass or weight, which was the usual practice in the larger houses of a hundred years ago or more. In such houses it was usual to interpose a 50–75 mm (2–3 inch) layer (pugging) of hair lime mortar between the floorboards and the ceilings underneath. An alternative to hair lime mortar was a 75 mm (3 inch) thick layer of sand. It will be appreciated that a 75 mm (3 inch) layer of sand can add a substantial dead load of about 200 kg per sq. metre (40 lb per sq. ft) to the floor, and few first floor joists were of sufficient section to support such an additional load for long periods. In addition to a sound-absorbing material between the floorboards and the ceilings, the laying of an independent floor on top of the first floor is a further benefit to reduce sound transmission. Further guidance on improving sound insulation is given in *BRE Digest 266* 'Sound insulation of party floors' and *BRE Digest 293* 'Improving the sound insulation of separating-walls and floors'.

5 Walls

External walls may be of stone, flints, cob walling, brickwork, half-timber, weather-boarding, vertical tile hanging or shingles. Solid brickwork can be rendered externally with a lime and sand mix, or, more recently, with a cement, lime and sand mix. If trowelled smooth, whether marked in courses to simulate stone walling or left plain, the finish is usually called *stucco*. In smaller domestic work, pebbles are frequently incorporated in the second coat of rendering. The pebbles may be mixed in the second coat before it is applied, when the finish is called *rough cast* or *wet dash* finish. Alternatively, the pebbles may be pressed into the second coat after it has been applied and before it has set, this is *pebble dash rendering*.

Stone walls

In many parts of the UK and abroad, stone has been widely used as a building material but, except in repairs and for more important buildings, it has largely been superseded by brick for reasons of cost. Building stones are naturally formed rocks, quarried or mined to provide building material. There are four basic types: granite, marble, sandstone and limestone. These stones are composed of various minerals which are inorganic substances, crystalline in structure and of different definable chemical composition. The age of these rocks is measurable in hundreds of thousands or millions of years, but once quarried and exposed to the air, they begin to deteriorate. The exposed faces can be worn away by the effect of wind, rain, high and low temperatures, sulphur and other gaseous impurities in the atmosphere of urban and industrial areas. The effective life of some may be less than 100 years, whereas granites and several of the harder sandstones last some hundreds or even thousands of years, for example, the granite blocks of Stonehenge, the marble temples of ancient Greece, and the pyramids of Egypt.

Granites are course-grained rocks containing free quartz, the other main constituent minerals being feldspar and mica. Granites have a wide range of colours and the precise mineral content varies with the region. All are known for their durability. In the UK, granites are quarried in Devon and Cornwall, parts of Westmorland and in several areas in the Highlands of Scotland. Some granites can take a polish finish.

Marble is a stone in the metamorphic rock class for it has been subjected to incredible pressure and heat, causing a structural change in the material to take place. Marble is a hard and durable material but sometimes can be weakened by veins of coloured minerals which run through the stone. Marble takes an excellent self polish, similar to granite. However, it is attacked by acids and its finish can suffer in polluted atmospheres. Most marble is of foreign import from such places as Belgium, Italy, Greece, Spain and Portugal.

Sandstones are among the most durable building stones and consist chiefly of quartz grains (that is, pure sand) bound together with mineral solutions of dolomitic, ferruginous, or siliceous cements. Small amounts of other minerals, particularly iron compounds, give them their distinctive colour, which may be grey, green, yellow, brown or red. They are widely distributed in central and northern England, the borders of Wales and in Scotland. Probably the best known is York stone, which varies appreciably in colour, texture and strength. It is known for its hardness and resistance to crushing, which explains why it is virtually always specified for templates, padstones and bedstones.

Limestones contain mainly calcium carbonate, but they differ considerably in character and durability. Chalk, for example, is a limestone, being almost pure calcium carbonate. Limestones vary appreciably in hardness depending on geographical area. That in south-eastern England is soft and only used locally as a building material, whereas chalk from northern England, Scotland and Ireland is reasonably hard. Bath stone is relatively soft and although it hardens on exposure, most forms do not last well when exposed to urban conditions. Portland stone is also a limestone which, properly selected, is one of the most durable stones available.

There are a number of quarries producing granites, sandstones and limestones and these are listed with suppliers, fixers, cleaners and restorers in the *Handbook & Directory of Members of the Stone Federation*, (82 New Cavendish Street, London, W1M 8AD). The 1986 directory has 122 suppliers of limestones, 103 suppliers of sandstones, 26 suppliers of slate and 80 suppliers of marble and granite. In addition to these facts the handbook contains information on the use of sealants in stone buildings, guidance notes on the use of stone, standard specification clauses and a useful bibliography of publications on natural stone.

Identification of all the varieties of stone is beyond the scope of architects or surveyors, although those who practise in districts where the local stone is extensively used will no doubt recognise the two or three characteristic varieties. Even in a reconnaissance survey of a stone building, the stone should be named and an indication given of its wearing qualities. If the surveyor is not a specialist, he should obtain his information from a competent firm of stone-masons.

Stone walls may be built in courses with all the stones in any one course of the same height but not necessarily of the same length. Also, stones may be in random sizes, either square, rectangular or irregular in shape. It is not unusual to find occasional stones equal in height to two or even three courses—these are called jumpers. The exposed faces may be rock faced, which is the natural surface of the stone as found in the quarry, but freshly fractured, when it is supplied as random rubble. Alternatively, the exposed face may be worked with mason's tools to give a variety of finishes. Stone finished to a smooth surface is known as *ashlar*, whether built in courses or as random work. Some examples of stone walling are shown in figure 10.

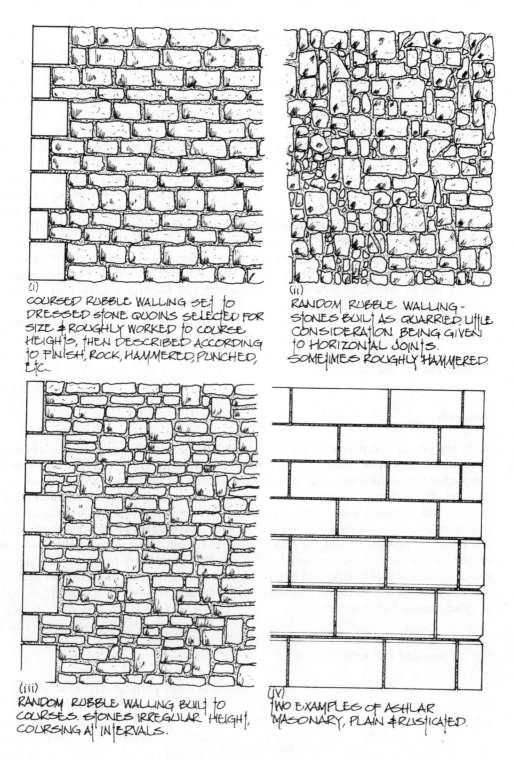

(i) COURSED RUBBLE WALLING SET TO DRESSED STONE QUOINS SELECTED FOR SIZE & ROUGHLY WORKED TO COURSE HEIGHTS, THEN DESCRIBED ACCORDING TO FINISH, ROCK, HAMMERED, PUNCHED, ETC...

(ii) RANDOM RUBBLE WALLING - STONES BUILT AS QUARRIED, LITTLE CONSIDERATION BEING GIVEN TO HORIZONTAL JOINTS. SOMETIMES ROUGHLY HAMMERED.

(iii) RANDOM RUBBLE WALLING BUILT TO COURSES. STONES IRREGULAR HEIGHT, COURSING AT INTERVALS.

(iv) TWO EXAMPLES OF ASHLAR MASONARY, PLAIN & RUSTICATED.

Figure 10 Stone walling

Only a small proportion of stone external walls are likely to be built in solid work, bonded from face to face. This is particularly true of thick walls, 750–900 mm (2 ft 6 inches to 3 ft), of older buildings. Behind the face, whether rubble or ashlar, there is likely to be a core, often of inferior material. Modern stone buildings usually have no more than a cladding of stone 100–150 mm (4–6 inches) in thickness, backed by brickwork or other building materials. The surveyor can rarely establish what is behind the stone face in other than modern buildings with joist floors. If there is room to crawl under the suspended ground floor it may be possible to see the interior faces of the external wall, then at air vents positions, the core material may be identified.

When reporting on the condition of the wall, the state of the pointing is very important. In recommending cleaning an old stone wall, washing with water should be discouraged. This is because much damage to the internal face of thick walls and to timber housed in such walls has been caused by excessive use of water, often before open joints have been re-pointed. An entirely satisfactory finish can be achieved on rubble walls without using water. The loose face must be removed by scaling off and brushing down. This will reveal any small areas or single perished stones that have to be cut out and replaced by matching, irregular-shaped stones. Final cleaning down, including removal of blackened areas, is done by lightly tooling the face of the stones with a mason's hammer. For re-pointing, a mix of 1 part of cement to 3 parts of hydrated lime and 12 parts of stone dust is suitable for ashlar, dressed stone and all stonework with thin bed joints. In random rubble walls the joints are often wide, and frequent re-pointing may have resulted in the arrises of the stones becoming lost. The Society for the Protection of Ancient Buildings (SPAB) at 37 Spital Square, London E1 6DY, has produced a great deal of literature on treatment and remedial work to old stone walls.

Flint walls

Flints are one of the purest forms of silica, which occur in veins in certain geological formations, particularly natural chalk. They are typically grey in colour and covered with incrustations. Although drab in appearance, they are extremely hard and resistant to exposure. Where available locally, flints were extensively used for cottages, farm houses, churches and boundary walls. They were set in mortar, usually with quoins, and dressings to window and door openings in stone or brickwork. Building with a random-sized material is a laborious process, calling for true craftsmanship but, when wages were low, it was economical to use the local material. With the development of transport facilities and low-cost machine-made bricks and concrete blocks, the advantages are no longer sufficient to offset the high cost of labour and materials. Flints are little used today, except in repairs or enlargement of existing buildings.

Pisé de terre and cob walls

Soils with cohesive qualities were extensively used in many parts of the world for walling. Basically, *pisé* construction is well-rammed soil containing several constituents, including a high sand content, sufficient clay to bind the soil and improve compressibility, and then enough silt to produce a rough grading of material. Various construction methods were used, for example, sometimes the material was consolidated *in situ* by ramming using timber shuttering, and sometimes it was made into blocks and built as brickwork. *Pisé* construction, incorporating modern improvements, will continue to have severe limitations, such as the necessity for craftsman to work the material, the need for a generous eaves overhang to provide protection from the rain and, finally, the demand for proper maintenance in respect of rainwater disposal arrangements. In the UK it should be regarded as suitable only for sheds, garden houses, apple stores, and the like.

Cob walling, associated with certain parts of the UK such as Kent, for example, involves differing techniques. Materials are ground chalk or clay and straw mixed with water, compacted into a soft but coherent mass. The cob wall requires a footing wall of brick or rubble, 450–600 mm (18–24 inches) wide, raised 225 mm (9 inches) above ground level. It is then raised in courses about 300 mm (12 inches) high.

Brick walls

Bricks have been used for hundreds of years in all parts of the world where suitable soils exist for their manufacture. The raw material clay or shale consists mainly of aluminium silicate, derived mostly from the decomposition of felspathic rocks. Natural clay deposits vary in chemical composition, depending on the proportions of their different mineral constituents. Deleterious salts in some clays, particularly gypsum, iron sulphide and magnesium sulphate, affect the durability of bricks when exposed to the weather. There are very considerable differences in properties of bricks from different types of clay, for example, the extremely hard Staffordshire blue engineering bricks, and, say, the flettons made from Oxford clay. A wide variation also occurs in bricks from the same district, or even from a single batch, depending on the thoroughness of the firing.

Bricks have also been made from mixtures of sand and lime which, after pressing into the required shape, are heated in steam to give them strength. These sand–lime (calcium silicate) bricks are produced in qualities suitable for internal, external, and special-purpose walls. Their resistance to damage by frost is related closely to their strength. They do not cause efflorescence trouble, but suffer a larger drying shrinkage than do clay bricks and this can cause cracking in large panels of brickwork. If the mortar is weaker than the bricks, the cracks follow the mortar joints, whereas with strong mortar, the cracks may run through bricks and joints indiscriminately. It is important when investigating cracks in walls built of sand–lime bricks to recognise the cracking pattern associated with such bricks. It is often the case that cracks are attributed to foundation movement, when in fact the amount, size, disposition and the age of the cracks together with the brick and mortar type and the size of the brick panel investigated, clearly indicate that shrinkage of the walling material is the true cause.

Concrete bricks or blocks are also widely used in many parts of the UK. These are made from Portland cement and sand and have the same drying shrinkage characteristics as sand–lime bricks. Some concrete blocks are made with dense aggregate, and others with light-weight aggregate, or in an aerated form. Special concrete blocks made with natural stone aggregate in the facing layer are known as 'cast-stone'. The best of these may, at a distance, be mistaken for natural stone. They share with concrete bricks the tendency to promote cracking at the mortar joints.

Bricks are available in many shades of white, grey, yellow, brown, red, green, blue, purple, black, and multi-coloured, and in a variety of textures. A surveyor cannot be expected to identify the many different kinds of bricks and, unless certain, he would be wise not to name them. What is important is the condition of the brickwork and the state of the pointing. He should particularly note the condition of bricks below the damp-proof course, or if there are none, those at ground level and in exposed positions, for example, in parapet walls and chimney stacks. It is advisable to test bricks in such positions with the point of a sharp tool to establish that the surface is hard. Flaking or spalling calls for further investigation. It is unwise to diagnose the cause of such defects from visual inspection alone, for the trouble may lie in the inherent characteristics of the brick. The smooth-faced stridently red, soft facing bricks so extensively used in the Victorian suburbs for example, are particularly prone to flaking. Spalling may stem from the inclusion of a small proportion of under-burned bricks, or it may be the result of frost damage. If only occasional bricks are affected, they may be cut out, but their replacement can present considerable difficulties if they are hand-made and from a brickyard no longer in production. Matching is then a real problem, although a good builder's merchant can usually find a suitable replacement or may know of a source of sound, second-hand matching bricks.

If there are large areas of defective brickwork, the surveyor must be extremely cautious about remedial measures. If failure is the result of rising dampness, re-facing without eliminating the root cause will not effect a long-term cure. If, however, the trouble stems from the exposed nature of the site or the porous nature of the bricks, re-facing the bad areas followed by rendering the whole wall may provide a lasting solution. It is important in recommending rendering to specify a suitable mix, which should be weaker than the background to which it is to be applied, see *BRE Digest 196* 'External rendered finishes'. If it appears that the trouble stems from soluble salts in the bricks, chemical analysis is necessary before deciding the appropriate remedial measures.

Defective pointing calls for careful examination. If the cause is deterioration of lime mortar from age, it is important to stress that the mix for re-pointing must be as weak. Now that the production of Lias lime has ceased, it is necessary to use a small quantity of cement in the mix. When recommending re-pointing, advice about the mix should be given because clients are apt to regard re-pointing as a simple maintenance item that can be left to the local builder, without professional guidance. For old lime mortar brickwork, a mix of 1 part of cement to 2 of lime and 8 of washed sand would be appropriate, whereas for brickwork in cement mortar a mix of 1:1: 4–6 would be more appropriate. *BRE Digest 160* 'Mortars for brickwork' gives helpful advice on jointing and pointing mortars. Particular attention should be paid to small areas of re-pointing such as behind downpipes, or making good in patches. Such re-pointing frequently appears sound at first sight and if tested, the face may be found to be hard, but it can sometimes easily be dug out in strips because of shrinkage of the mortar in the joint. The trouble usually stems from using too rich a mix, but can result from failure to rake out the joints sufficiently first.

For domestic work today, cavity wall construction is used where the wall is built in two leaves, with a 50–75 mm (2–3 inches) cavity between the leaves. In more substantial older buildings of more than two floors, the inner leaf is often a full brick, or 215 mm (9 inches) in width. Cavity wall construction can almost invariably be detected because the facing work consists only of stretchers, but the thickness of the wall should be checked at window openings to establish that the construction is true cavity work. In modern estate development, the inner leaf is often built in timber, breeze or cellular concrete blocks which possess good thermal insulation qualities. Cellular concrete blocks have the disadvantage of shrinking appreciably as the building dries out, and plaster cracks are often a considerable problem.

When cavity wall construction was first used, the spacing of ties was apt to be rather poor and, therefore, such walls may be far from stable. Severe corrosion of a number of steel wall ties has also occurred causing extensive failure of several cavity walls, and this is covered later in the chapter.

In addition to these cavity wall problems, where the wall plate is bedded on the outer leaf of the wall and the feet of the rafters are not adequately tied, the thrust of the roof may result in the top of the cavity wall tilting outwards, making the roof and the top floor unstable if not actually dangerous.

Features for further investigation

Brick and masonry walls frequently reveal the changes a building has undergone. For example, because of 'Window Tax' in olden days, it is not uncommon to see the outline of a window with the frame and glass area built up in brickwork or stone. Window openings may have become redundant for many other reasons and the old craftsman were often content to build up the void without removing lintels or arches, jambs or sills. This has sometimes resulted in old timber lintels or frames being sealed in, giving rise years later to fungal decay or beetle infestation.

Extensions tended to be butted to the original work, instead of being properly bonded in, and the resultant straight joints provide ready access for water. The positions of built-up voids and straight joints should be noted for later careful examination of the internal faces. Some alterations call for re-distribution of loads, and this problem did not always receive the necessary attention with the result that settlements have subsequently occurred. Such settlements require the fullest investigation. Scoured surfaces behind downpipes, algal growths and areas of defective pointing should be recorded and any timbers housed in walls in the vicinity investigated during the internal survey. Projecting string courses should be viewed with suspicion since, with age, the weathered top edge may be worn away, resulting in water being led into the wall. Balconies are another feature that warrant close attention. Cracked sills may be particularly insidious in older buildings where large-sized timbers have probably been used as grounds for window boards or panelling under windows. In one case, a severe outbreak of dry rot was traced to defective external sills on the second floor of a house for there was extensive weevil attack in the grounds, although this wood-boring weevil is usually only found in decayed basement or ground floor timbers, and it is not known to be capable of flight.

Weatherboarding and cladding

In some parts of the UK, particularly parts of Surrey and Kent, the external walls of both cottages and more important buildings are often of weatherboarding on studding, the internal faces being lath and plaster. If the buildings are old there is usually no thermal insulation between the weatherboarding and the internal plaster face and no felt behind the weatherboarding, unless it has been renewed *in toto* within the last fifty years or so. If original, the weatherboarding will almost invariably be feather-edged European redwood, usually painted.

In barns and in buildings where only part of the wall is of weatherboarding (usually gable ends), the boarding may be random-width elm boards. Provided such buildings have been well-maintained, including regular repainting and upkeep of gutters and downpipes, they are often in remarkably good condition, largely because the softwood timber available prior to the First World War was relatively free from sapwood, and hence will have escaped serious furniture beetle attack. Inspection of these weatherboarded buildings presents very real difficulties because the vital structural members (the studs) are hidden. In a building one hundred years old or more it is most unlikely that there will be any continuing active furniture beetle attack unless new timber has recently been introduced in repair, and old flight holes are likely to have been concealed by frequent re-painting. When inspecting such buildings the surveyor has very little to go on in assessing the soundness of the hidden timbers. This makes it imperative to inspect roof voids thoroughly, and to take up floorboards against external walls, as the condition and quality of roof timbers and joists will provide some indication as to the probable condition of the studding.

When recommending thermal insulation to weatherboarded houses, changing domestic habits should not be overlooked. Today, much more water vapour is produced within the house from frequent baths, the use of tumble driers and similar appliances. This factor could lead to condensation occurring within the studding which will be a problem. For kitchens and bathrooms, it is advisable to recommend applying a vapour barrier on the warm side of the wall in addition to the thermal insulation. A suitable finish to insulation board which has been fixed to the wall would be foil-backed plasterboard.

In the USA and Canada most houses outside town centres are still timber-framed with timber cladding, that is weatherboarding or shingles. Houses and bungalows built to Canadian standard in this country are usually entirely satisfactory. Unfortunately, weatherboarding has been used here for external sheathing without basic principles being properly understood at the time, particularly in houses and blocks of flats where it was used as an architectural feature, either from first floor joist level to the eaves, or in panels, usually under windows. The cladding has been applied on battens to breeze concrete or cellular concrete walls without a vapour barrier on the warm face of the wall. In addition, instead of a breather-type paper behind the battens, a common mistake was to use a bituminous felt paper or, even worse, an aluminium-backed felt, fixed between the battens and the back of the boards. This effectively retarded drying out when water penetrated the face of the boarding. Moreover, instead of traditional overlapping weatherboarding, V-jointed, tongued and grooved boards were widely used. This was and still is permissible in the USA and Canada, where the cladding is western red cedar or a similar decay-resistant softwood and only breather-type papers were used. In the UK the timber was more often European redwood containing an appreciable amount of sapwood, which was not at all resistant to decay. Further, at this time, it was exceptional for the timber to be pressure-treated or kiln-dried to an appropriate 15 per cent moisture content. In consequence, early failure of such cladding was all too common.

As cladding is not a traditional method of construction in this country, other mistakes occurred. Inferior pink primers were commonly used. These were rarely applied to top and bottom edges and cross-cut ends, but sometimes to the back of the boards, as well as the face, thus reducing the chances of escape of the excessive moisture originally in the timber. This often resulted in early paint failures or fungal infection. Galvanised rather than aluminium nails were often used, or even ordinary wire cuts which produced early rust staining. Nailing technique was often wrong, encouraging splitting of boards and loss of support. Tongued and grooved boards have been known to be fixed upside down, that is, with the groove for the tongue upwards, with the result that water ran down the face of the board and into the groove for the tongue. European redwood, unless pressure-treated and kiln-dried, should not be used as horizontal tongued and grooved boards. In preference, western red cedar, kiln-dried to 15 per cent moisture content, followed by a deluge treatment with a 5 per cent solution of pentachlorophenol or similar, should be used. The boards should then be stacked between battens to allow the carrier solvent to evaporate before being primed on face and on the top and bottom edge and crosscut ends. One undercoat of paint on the primed surfaces prior to fixing is recommended. In addition to using a breather-type paper behind the pressure-treated battens, and aluminium nails for fixing, provision of adequate ventilation behind the boarding must be provided.

V-jointed timber boarding on new property should be viewed with suspicion, particularly if it is painted, indicating that the timber is almost certainly European redwood and not western red cedar. Unless the surveyor is satisfied that the faults described above have been avoided, and that the boarding, if European redwood, has at least received a dipping treatment with an appropriate fungicide, he must stress that the life of the panels could be very short. Moreover, if the V-jointed panelling is open to criticism, there may also be faults in assembly of the external joinery. The surveyor should advise against purchase of a property if he is not entirely satisfied with the V-jointed cladding, because, having found a failure to appreciate the significance of certain vital design details, there may well be others that he is in no position to discover, for example, an unventilated flat roof void, constructed of inadequately treated timbers, and possibly with no vapour barrier.

Where combustible materials are permitted for external wall surfaces, plastic as an external cladding material is becoming more prevalent, particularly for the modern timber-framed house. Extruded rigid PVC sections in sheet or narrow board form has increasingly replaced traditional timber cladding, because it is said to be maintenance free and relatively cheap. The surveyor should inspect the fixings of such claddings as these can fail owing to the high rate of thermal expansion of the PVC. Cracks may be evident around fixings where supporting nails and screws have been driven home without allowing for movement to take place via the preformed slots within the sections. In addition during cold weather, PVC cladding will become brittle and thus more prone to impact damage.

Attention should be drawn in the survey report to the high cost of insurance of buildings and their contents when the building is constructed of combustible material, that is, those in other than brickwork or masonry.

Shingles

Wooden shingles involve the same principles as vertical or horizontal cladding. It is to be presumed that the shingles will be 'rift-sawn' western red cedar, which is a very durable timber, and they are usually fixed to horizontal or diagonal boarding. The Canadian Authorities approve of the shingles being nailed direct to the sheathing, but there must be a vapour barrier on the warm side of the wall, with the thermal insulation the vapour barrier or between the studs.

Vertical tile and slate hanging

This treatment has been extensively used in some parts of the UK in both old and modern housing. In old houses the tiles will be hand-made clay tiles, and the slates those commonly used in the district as a roof covering. These are fixed to horizontal battens nailed to stud walls. Before bituminous felt papers became available, no provision was made to prevent penetration of water into the studding should tiles become dislodged. For insulation it was not unusual to pack hay behind the tiles. Such buildings are as difficult to survey, if not more so, as those with external weatherboarding, and for the same reason, the studding is inaccessible. The surveyor must try to ascertain whether the tiles or slates are fixed with timber pegs or nails and if these are still sound. The tiles or slates must be even more carefully scrutinised than those used as a roof covering because the backs cannot be seen. It is important to look for evidence of patching. If there is extensive algal growth the battens may have been damp for long periods, and may be decayed. Unless there are gaps where tiles have slipped, allowing examination of the battens, the surveyor must either insist on lifting a few tiles or slates, or advise his client to budget for having to strip the covering and renew the battens at an early date. Improving the thermal insulation of such tile-hung external stud walls calls for similar precautions to those recommended for weatherboarding. Tile hanging, in particular, has continued to be used down to the present day, usually on only part of some elevations, for example, from first-floor joist level to the eaves, or as aprons under first-floor windows. The tiles of houses built prior to the Second World War are likely to be fixed to studding, but usually with an impervious bituminous paper behind the battens, and they require as careful an examination as in older houses. The tiles of houses built after the Second World War are likely to be fixed on battens to breeze concrete blocks, with the added hazard that the nails may become more rapidly corroded. Every effort must be made to establish how the tiles are fixed.

Timber-framed buildings

The stability of the original timber-framed building, with brick or wattel and daub infilling, is difficult to assess, since few of such framed walls are likely to be plumb and the significance of bulges is not easy to determine. Moreover, the method of construction, with the horizontal members housed in tenons at the ends of vertical posts, means that the original construction cannot be restored when the tenons decay, short of shoring up the roof and dismantling whole walls. Usually, recourse is had to iron straps to tie walls together. Examination of the external framework has to go hand in hand with the survey of the interior.

Outside, the condition of the sole plates and that of the ends of the vertical posts require careful inspection, and the bottom edges of the sole plates should be probed for decay which is likely to be of greater consequence than any death watch beetle infestation. If the death watch beetle attack is still active the surveyor can be certain that the attacked timbers contain appreciable fungal decay. Flight holes can be misleading, because many original period cottages and houses were constructed of salvaged timbers from earlier buildings, and the infestation may date from then. Had the attack continued for 200 or 300 years, collapse would certainly have occurred long ago. For recognition of death watch beetle, see appendix A. Woodworm (common furniture beetle) attack is unlikely to be significant, because it will almost certainly be old attack, unless new timbers, particularly of softwood, have been introduced in recent repairs. Probing quite heavily attacked surfaces may well produce a shower of frass (borer dust), but this is unlikely to be evidence of continuing active attack. It is much more important to examine the rims of flight holes with a pocket lens as sharp rims and light-coloured frass spilling out indicate continuing attack which, in sound as opposed to decayed wood, will be confined to the sapwood of oak and almost entirely to the sapwood of softwoods. A surveyor should not accept instructions to survey these traditional timber-framed buildings without some knowledge of timber identification. He must be able to distinguish softwoods from hardwoods and to identify the few hardwoods normally used for the framing members, such as oak, elm and sweet chestnut. Identification of the joists, particularly ceiling joists and rafters in smaller cottage properties, may present greater difficulties because locally available poles of a variety of species were often used.

Modern methods of timber-frame construction were introduced from North American and Scandinavian countries into the UK, in the 1960s. This type of construction consists of panels of timber studwork clad in masonry or composite external veneers. The external skins are not of themselves structurally significant, for the internal panels are self-supporting and normally take the roof load. These internal load-bearing timber panels are protected by timber sheathing, vapour barriers and breather papers.

Bad publicity in 1983 caused a sharp decline in timber-frame construction but recently numbers have increased as public confidence has returned. The cause of the down-turn in production was the publicising (perhaps overdone)of investigations which revealed a number of defects in the construction, these defects included:

(a) High moisture contents in studwork, particularly in members close to the floor plate.
(b) Missing or defective vapour and breather papers.
(c) Cold-bridging and condensation.
(d) Inadequate or missing cavity closers (these prevent the spread of fire).
(e) Missing or inadequate wall ties.

When faced with a timber-framed house, initial observation may not indicate the form of construction as the exterior may resemble the traditional brick and block dwelling. *Timber Research and Development Association (TRADA) information sheet No 10* 'Structural surveys of timber frame houses' lists a number of details which will help the surveyor to determine whether it is timber-frame construction or not. The information sheet is also a comprehensive guide for the surveyor as it highlights the critical areas of timber-frame construction which require particular attention when carrying out the survey. When reporting on such houses, the surveyor must inform his client of the problems of extending or altering the structure. And in order for the surveyor to advise on alterations and the like, he must be knowledgeable about the construction. *TRADA information sheets No. 3* 'Introduction to timber framed housing', and No. 5 : 'Timber framed housing—Specification notes' are useful texts on the subject for the surveyor.

Walls and floors

The surveyor will be concerned with internal partitions as well as the external walls. He will want to establish the thickness of partitions, and whether they are of masonry, brickwork, breeze, concrete or cellular–concrete blocks, or studding. He should ascertain which are load-bearing partitions and which are not, and whether the latter are built off the floorboards because this construction will add greatly to the cost of repairs should he have to advise taking up and re-laying such a floor. With stud partitions, and even more so if such partitions are true framed partitions incorporating some form of truss, the surveyor should be on the look-out for any alterations that may have been carried out. In many instances it is not unusual to find that some structurally vital truss member has been cut through to form a doorway in a new position. As long as the precise function of structural members is determined, it is perfectly feasible to cut off such an important one as a principal rafter, provided the load it is carrying is properly re-distributed.

Doorways out of square are an indication that settlements have occurred, but it is not sufficient merely to record the fact. Even in a reconnaissance survey the cause of the settlement must be established. For example, doorways to two facing bedrooms with a corridor between were found to be out of square. This defect had developed because the partitions on the ground-floor rooms beneath had been removed by a previous owner in order to make one large reception room. Thus the partitions on the first floor had been deprived of adequate support. In this case it was a matter of urgency to introduce steel beams under the partitions to prevent the collapse of the floors. More often, distortion of door openings can result from settlements in the foundations which occur from a variety of causes. Settlements in walls, and particularly of lintels or brickwork over door and window openings, may also be caused by decay in buried timbers such as bond timbers in walls and timber lintels over openings.

It is vital in a detailed structural survey to check that partitions on upper floors are immediately over partitions on the lower floors, although with large reception rooms on the ground floor, the internal partitions on the upper floors may well be carried on large beams in or just below the ceilings of the ground-floor rooms. The adequacy of such beams must, of course, be checked. Surprisingly often, in larger houses of some age, partitions on upper floors are not immediately above the partitions on floors below, sometimes being off-set only a few inches, which careful measurement of the rooms alone will reveal. In such cases, the problem of dividing up the accommodation differently on lower floors becomes more difficult.

Damp-proof courses and rising damp

The surveyor should check that damp-proof courses exist; they are by no means always easy to detect, for example, when they are obscured by mortar or by a cement and sand plinth at the base of walls. In this position such plinths may defeat the purpose of the damp-proof course by bridging it on the external face. In older dwellings, damp-proof courses were compulsory under the 1875 Public Health Act and these should be at least 75 mm (3 inches) above the external ground level, whereas current Building Regulations demand a minimum height of 150 mm (6 inches) and this is obviously preferable. Constant feeding of flower beds at the base of walls, or the laying of concrete paving after a house is built, not infrequently results in the horizontal damp-proof course being buried, and this is something the surveyor must not fail to observe and record. If ground levels outside are above floor levels inside, vertical damp-proof courses are as necessary as horizontal ones. The search for vertical damp-proof courses will usually necessitate excavating, and should be deferred until the condition of internal faces of the walls has been established.

Damp-proof courses serve a very necessary function although they are only found in more modern buildings. Many buildings without them are, however, basically sound, although there may be plaster problems in ground floor or basement rooms because of persistent rising damp. This rising soil moisture draws up salts from the soil, which work through to the face of the plaster. Being hygroscopic, the salts deposited on the face of the plaster attract moisture at times of above average humidity, discolouring the decorations. Persistent absorption of moisture by these salts, and subsequent drying out when the humidity falls, cause wallpapers to peel and paint finishes to blister. Rising damp usually

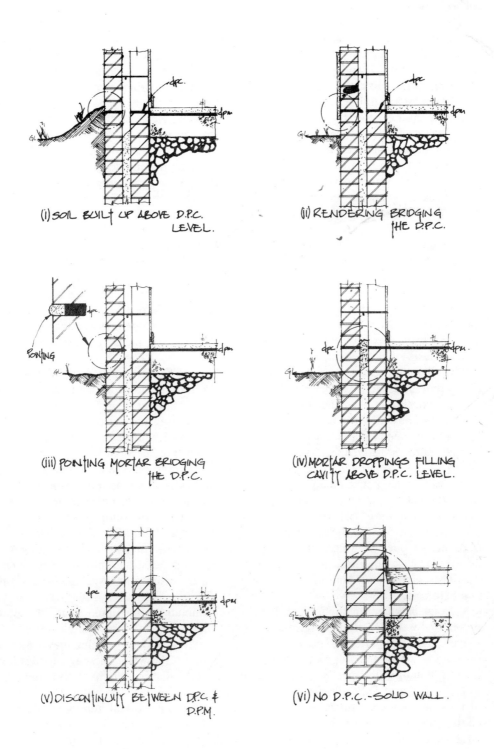

Figure 11 Typical causes of rising damp

only rises to about 1050 mm (3 ft 6 inches) above the external ground level, unless the walls are treated with a 'water-proof' paint externally or a 'water-proof' plaster internally, when it may travel up higher. Such treatments provide no permanent cure unless taken up from floor to ceiling. In most cases, the height attained by rising damp will depend on the pore structure of the wall and the rate of evaporation. In this way the internal environmental conditions will affect the symptoms and appearance of rising damp, for example, the installation of central heating or the fixing of double glazing. The changed conditions may lead the occupier to think that the problem of rising damp has been eliminated. *BRE Digest 245* 'Rising damp in walls: diagnosis and treatment' stresses the importance of distinguishing between rising damp and other causes of damp conditions. The length of time of the attack might be indicated by smell, deteriorating decorations, efflorescence and rot. If the problem is a recent one or the cause is intermittent, then these visible indications may not be present. The diagnosis, however, may not be conclusive, and the causes could be damp penetration due to defective construction, plumbing defects, condensation or hygroscopic salts.

In addition to spoiling decorations, rising moisture can cause decay in skirtings and panelling fixed to the wall, the decay usually being caused by a so-called wet rot rather than dry rot. Attack may spread behind door and window linings to considerably greater heights than moisture will rise in the solid walls. This will be due to the build-up in the relative humidity of the still air behind such linings, providing ideal conditions for fungal growth. Attack may even reach the head of a door opening although moisture in the solid walls has only risen to about 1050 mm (3 ft 6 inches). Figure 11 illustrates typical causes of rising damp.

When building costs were lower and the value of the property justified the outlay, it was not unusual to insert horizontal slate or lead damp-proof courses where none existed, working in short lengths of 600–900 mm (2–3 ft) at a time. Today, the process is much simplified by the use of hand or power-driven saws to cut the slot into which the damp-proof membrane is inserted. Modern damp-proof courses used in these situations are half-hard or soft copper, two layers of bitumen felt, or high/low-density polythene sheets, see *BRE Digest 77* 'Damp-proof courses'. The insertion of a horizontal damp-proof course eliminates rising damp, but does not cure salt contamination. Even hacking off contaminated plaster and re-plastering with special plasters may not completely eliminate salt contamination of brickwork, for some salts are likely to continue working their way through to the face of the new plaster, sometimes in as little as five years. One solution is to batten out the walls with pressure-treated battens, to which a dry lining should be fixed. This solution, however, may present numerous problems with openings, skirtings, architraves and cornice members, apart from upheaval to the occupants and high cost.

Various patented methods are claimed to eliminate rising damp, but for some of these, for example, electro-osmotic processes, there are no impartial scientific data to suggest that they effect a permanent cure. Several proprietary techniques which have proved successful aim to produce a moisture barrier in the wall. These techniques involve the injection of water-repellent substances in holes drilled at regular intervals along the base of the wall. Basically, these measures rely on silicone solutions with water-displacing fluids or siliconate–latex mixtures. As with the insertion of conventional horizontal damp-proof course, the boring and injecting of walls does not overcome the problem of salt contamination induced by prolonged rising damp which will have to be overcome as previously outlined for dry lining.

Ventilation

Timber floors in ground floor or basement rooms require to be ventilated, and air vents should be found in the perimeter walls below floor level. The surveyor should record the position of such vents in the external walls, particularly noting whether the vents are all of the same size and pattern. If they are of more than one size or pattern, suspicions should be aroused. They may indicate a previous history of fungal decay, evidence of which should certainly be searched for during the room-by-room survey of the basement and ground floor rooms. The surveyor should check that air vents are at least 75 mm (3 inches) above ground level, or that there is a sunken area, protected by a kerb, in front of them to ensure that water is not led through on to the oversite. He should also check that no air vents are obscured by dense vegetation on the external walls or by the soil having been raised above them by enthusiastic gardeners. It is not sufficient merely to observe that there are air vents, and in sufficient number. Each vent should be tested to establish that it is functional and not blocked by a wall plate or joists on the internal face of the wall; this is where the piece of stiff wire recommended earlier will be found invaluable. The surveyor must not be misled by air bricks that are blocked on the inside by concrete when an original joist floor has been replaced by a solid floor, without the air bricks being cut out and the wall built up solid. Apparently sufficient air vents are no guarantee that ground or basement floors are adequately ventilated beneath. There may be areas of solid floors interrupting through or cross-ventilation and, in larger buildings, it is unlikely that air vents alone will secure sufficient under-floor ventilation unless there is an impervious membrane in, under, or on top of the oversite concrete. Such a membrane will eliminate rising soil moisture and so reduce the likelihood of high relative humidities being built up under the floor. Provision of an impervious membrane in the oversite concrete makes it all the more important to ensure that water is not led through the walls on to the oversite, where it cannot escape by seeping through the concrete into the hardcore underneath.

The subsoil is often all-important, for as previously mentioned, where it is gravel, unventilated basement floors with no oversite concrete have been found to be still sound, eighty years later. On the other hand, gravel provides an easy entrance for water and in low-lying areas close to rivers, the water may on occasions rise above the gravel oversite.

Ground levels

The relationship between walls and the ground levels, outside and inside, is particularly important and should be noted on the survey report. If the ground level outside is above floor level inside, then there should be both a vertical and horizontal damp-proof course to provide a continuous barrier to moisture. Ground or basement floors in such cases are unlikely to be adequately ventilated and, unless they are solid, the possibility of appreciable fungal decay in the floor timbers should be suspected.Ground levels outside may have an important bearing on possible flooding, particularly if there are watercourses or streams in the vicinity. It may be necessary to check with Local Authorities or Water Authorities to establish whether or not a risk exists. It is not uncommon for apparently insignificant streams to take very large quantities of water at times of abnormally heavy rain. The danger from flooding is often increased in residential areas if such streams are taken under roads in culverts. Failure to keep these culverts clear can result in the water rushing over the road,

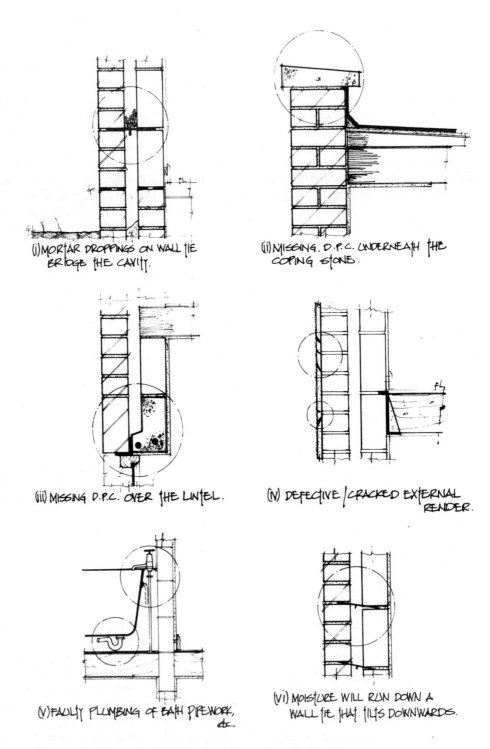

(i) MORTAR DROPPINGS ON WALL TIE BRIDGE THE CAVITY.

(ii) MISSING. D.P.C. UNDERNEATH THE COPING STONE.

(iii) MISSING D.P.C. OVER THE LINTEL.

(iv) DEFECTIVE/CRACKED EXTERNAL RENDER.

(v) FAULTY PLUMBING OF BATH PIPEWORK, etc.

(vi) MOISTURE WILL RUN DOWN A WALL TIE THAT TILTS DOWNWARDS.

Figure 12 Typical causes of penetrating damp

flooding property that might otherwise escape were the flood water to have kept to the course of the stream. Where the ground slopes towards an external wall, the position of air vents is important. If these are close to ground level, particularly if there is paving or concrete against the external wall, there is a considerable risk that water may be led through on to the oversite concrete. Water laying on the oversite is very liable to produce wet rot in the floor joists and flooring even when there is an appreciable distance, for example, 600 mm (2ft) between the oversite and the underside of the floor timbers.

Penetrating damp

Rain penetration will be established by the type, orientation and condition of the wall and the pattern of internal wetting, for example, following a period of prolonged and heavy rain. Older buildings with solid walls of poor quality (under-fired) bricks are very susceptible to penetrating damp. Poor jointing and pointing will also increase the moisture penetration of brickwork and stonework. Often the cause of penetrating damp will be a combination of a number of such factors which together create the conditions suitable for the defect to occur. Figure 12 illustrates typical causes of penetrating damp.

The surveyor must not miss the obvious when finding dampness on the inside wall of a house; for example, leaking water pipes and defective roof coverings are the most common instigators of damp penetration. In addition, broken downpipes and gutters, poorly connected plumbing equipment and leaking domestic appliances are all potential sources of moisture which can penetrate masonry. Plumbing in old properties is usually a confusing tangle of pipes some of which may be buried within the walls, and old lead pipes are frequent sources of minor leaks. Short overflow pipes can also create damp conditions on windward elevations and therefore should be checked.

The surveyor must be able to differentiate between condensation and penetrating damp, for a wrong deduction can lead to unnecessary expenditure, loss of confidence and continuation of the problem.

The remedial work to be recommended for penetrating damp can only be specified with confidence after exhaustive investigation. The cure may be one of the following:

(a) Clearing obstructions in the cavity—if cavity walling.
(b) A silicone water repellent treatment to the external surface (a short-term solution).
(c) The application of rendering or cladding.
(d) Cutting out of porous bricks or stone and replacement with more durable materials.
(e) Re-pointing with a suitable mortar.
(f) Injecting further insulation, if the cavity is insulated.
(g) Demolition and removal of faulty cavity insulation and proper replacement.

With regard to cavity insulation, the effect of the inserted insulation will be to raise the moisture content of the outer skin and to lower its temperature. These levels may be very marginal and have little effect unless the outer skin is already deteriorating, in which case an increased rate of deterioration might be anticipated.

Condensation

Condensation is a major problem with domestic properties built prior to the Second World War, and with a number of high-rise concrete dwellings. It has also appeared in a number of low-rise timber-framed houses, which were built in the 1960s and 1970s. Its occurrence can be attributed to changing use of buildings, increased and fluctuating heating levels, restricted ventilation by sealing off open fireplaces, draught proofing and the replacement of poorly fitted windows and doors, and inappropriate design and poor workmanship.

Condensation can affect walls in two important ways. Firstly, it can create the environment which promotes the decay of structural timbers by dry/wet rot, as well as beetle infestation. Secondly, condensation (or to be more concise *interstitial condensation*) can increase saturation levels within the wall, leading to timber decay in panel walls and, over time, slow degradation of masonry walls, particularly the jointing material. Condensation, its causes and remedies is covered in detail in *BRE Digest 110* 'Condensation' and in *Digest 297* 'Surface condensation and mould growth in traditionally-built dwellings'.

Defective rainwater goods

Improper maintenance, poor design and bad workmanship are reasons why rainwater goods, such as gutters, downpipes and gullies, fail. Some of the most common problems are blocked, cracked, broken and missing gutters and downpipes. These defects can lead to water penetration of masonry at all levels. Evidence of such defects in rainwater goods is best obtained during periods of rainfall. Alternatively, testing can be carried out by playing a hose pipe across the roof and observing the effect. Leaks will become evident as well as damp staining of the masonry.

The backs of downpipes should be examined. Most often they will not have been painted for many years with the result that an apparently sound and well-maintained downpipe is actually badly corroded or paper-thin at the back, if not completely split. The nature of fixings for pipe nails should be noted and, if of timber, whether these fixings are still sound. Walls should be examined for evidence of gutters and pipe runs having been re-routed, often without removal of the old timber fixing plugs. Thus there may well be decay in timbers housed in walls in the vicinity of these former runs.

Cooling flue gases

Water-vapour, when produced in large quantities in domestic flues, can lead to condensation penetrating the chimney walls. In addition, salts and other products of combustion can be carried by this moisture through the wall to appear as staining on plaster and decorations. The risk of condensation occurring is greatest when coke, anthracite and similar slow-burning fuels are burned in enclosed spaces such as boilers and continuous-burning grates, but not with coal burned in open fireplaces. One reason why condensation occurs with such appliances is that the temperature of the flue gases from slow-burning fuels is lower at the top of the flue. Also, these gases are not drawn up the chimney by the fast moving air currents that are so prevalent with open fires.

With the change over to electric fires and central heating, many fireplaces are sealed, but not their flues. This can result in rain penetrating the flue and, with no heat to dry off this rain, damp staining, discoloured by deposits on the walls of the flue, may appear on plastered breasts of rooms on the top floor. Such stains are very likely to reappear after redecorating unless the old plaster is completely removed. The stains are nearly always caused by contamination with chlorides (common salt) derived from the soot. These chlorides enable the plaster to extract moisture from the air at times of above-average humidity, producing discolouration of the decorations on the chimney breast. The contaminated plaster must be removed and the mortar joints of the brickwork raked out squarely to give a good key. The brickwork can then be rendered with a mix of 1 part of Portland cement to 3 parts of clean sand, keyed to receive a finish of gypsum plaster. Any areas of contaminated ceiling should be replaced, preferably with a vapour barrier such as sheet polythene separating such areas from the wall. Having dealt with the problem of discoloured plaster, disused flues should be sealed if they are not to be used for ventilating rooms. To seal flues, chimney pots and flaunching to the pots should be removed and replaced with a reconstituted stone coping bedded on a damp-proof course on top of the stack. When flues are no longer required it is preferable to take down disused stacks to below the roof line, making out the roof with slates or tiles to match the remainder of the roof covering. This disposes of the problem of future maintenance of the stack and its flashings. Flues are, however, a useful means of ventilating rooms and can assist in minimising possible condensation problems. The course to be adopted in these circumstances is to seal off the fireplace openings, inserting a plastic louvred vent in each flue just below the ceiling level. The stack is dealt with by removing the chimney pots and flaunching, capping the stack with a coping, bedded on a damp-proof course, and inserting one air brick in one side of the stack in each flue.

Boundaries

Boundary walls must be inspected for condition and structural integrity. Being free-standing, without piers and often 2 metres (6 ft 6 inches) or more in height, support may be found lacking, particularly if the wall is only 215 mm (one brick) thick. The highest part of the wall should also be closely inspected, for in many cases deterioration will be advanced because of inadequate or missing copings.

Boundaries may complicate maintenance, for example, if an external wall is so close to the boundary, there may be insufficient room to erect ladders to paint gutters, downpipes and external joinery. Also, it will be difficult to erect scaffolding to permit repairs to roof slopes or re-pointing of walls, without trespassing on to a neighbour's land. Thus, it may be impossible to carry out essential maintenance and the surveyor should be aware of this when inspecting the house. Except in leasehold property where one ground landlord may own whole streets, it is most unlikely that there is provision in the deeds of a property for an owner to go on to his neighbour's land, to repair his own property. In these cases, the owner would be dependent on the goodwill of his neighbour. Often, however, an existing owner has an understanding with his neighbour and permission is readily obtained. Such gentlemen's agreements, however, have no force in law. Moreover, there is the possibility that the adjoining property may change hands and the new owner may be unwilling to grant concessions. The surveyor should certainly draw attention in his report to such boundary complications where they exist and not leave such warning to the solicitor dealing with the conveyancing side of the transaction.

Party walls of semi-detached and terrace properties can also give rise to difficulties. In practice, there is joint responsibility for defects in party walls, or chimney stacks straddling a party wall, but it may not always be easy to agree that repairs are essential, and the adjoining owner may just not have the funds to discharge his share of the joint liability. Other complications can arise, for example, in small semi-detached properties, water from certain roof slopes is often taken away by a downpipe attached wholly to the adjoining owner's property. An owner has no right to go on his neighbour's land to repair this pipe, yet its continuing want of maintenance may cause an outbreak of dry rot in his property.

Finally, boundaries may be important because of the possible consequences of road-widening schemes, or the construction of new roads, or the future development of adjacent land which could spoil the existing amenities. The surveyor carrying out a reconnaissance survey should certainly report on these aspects in his report, stating where enquiries should be made if, in fact, he does not make them himself. The County Highway Authority will be able to advise on road-widening projects or new roads, if these have already been decided upon. The planning officers of the Local Authority will provide information as to future development plans where these have been determined.

Close proximity of adjacent buildings may affect intended alterations, even where there is sufficient room to do the intended work. For example, the owner of a new building has no right of light from his neighbour's land. Moreover, if a building owner forms new windows in a flank wall and the neighbour then erects a high boundary on his land, thereby obstructing light, the building owner has no redress. There is also no right of privacy, and an owner cannot restrain his neighbour from forming a window which overlooks his own property, although planners may take a different view.

Cracking

When cracking occurs in walls, a client will want to know quite early on whether the failure is serious and if collapse is imminent, or whether some reduction in performance is likely. Frequently a state of anxiety has been introduced by external sources, for example, the press and television reporting on extreme weather conditions or by inadvertent conversation with those having little technical knowledge, which is difficult to alleviate or qualify later.

It will have been seen from previous chapters that there is seldom only a single cause for a failure in an element or component and that, frequently, the failure is only a nuisance initially which may go unnoticed for some time and occasionally for too long. Some failures are more likely to be observed than others of course, especially by those who clean internally. For example, cracking in plastering is usually quickly noticed and given a high anxiety rating, but cracking in brickwork would go for some time without comment, even if it should be noticed. On many occasions, only when attention is drawn to the crack do occupiers seek advice. Often by then, it is usual to find that the crack has been categorised as a serious failure, and by this time it is difficult to reassure a client, when such a degree of seriousness has been assumed for some time.

When inspecting walls which have crack-patterns showing, it is important to appreciate the following factors pertaining to the wall:

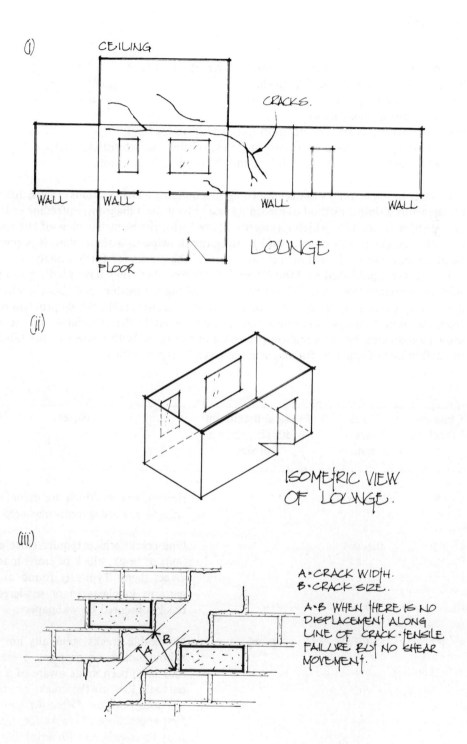

Figure 13 Recording and measuring cracks

(a) length;
(b) height;
(c) thickness;
(d) original specification, bond, mortar designation, etc.;
(e) original design, materials, finish, etc.;
(f) perforation pattern;
(g) subsequent maintenance;
(h) any alterations since completion (physical or in loading);
(i) changing or changed environment (trees, subsoil characteristics, etc.);
(j) physical life cycle of the building and the element in question.

In order to understand what is happening to a building which contains cracks, John Pryke recommends a simple method of recording cracks by using a diagram, unfolding every room as shown in figure 13(i) which represents figure 13(ii), the isometric view of the room. By recording cracks in this manner and relating the drawing to a floor plan, it is possible to build up an accurate three-dimensional record of the cracks for further analysis.

From a table published by John Pryke & Partners, which they have kindly given permission to reproduce here, a definition scale of cracking is possible, and this classification is a most important stage in the inspection procedure. However, the whole problem resulting from this defect can be very subjective and consequently the application of the analysis must be tempered by the situation presented at the time. With respect to the table below the definitions of crack width and size are given in figure 13(iii).

Class of crack	Crack size mm	Physical maximum width (B), shown full size	Notes
P0	Less than 0.1		Hairline cracks which are difficult to see unless walls are smooth plastered
P1	0.1–0.3		Fine cracks which require close observations of brick, block or stone masonry to detect them. Typically found as crazing patterns in plastered or rendered walls. Easily obscured by wallpaper
P2	0.3–1.0	▶‖◀	Noticeable cracks, generally not seen by a householder or property owner unless they have been made aware of a possible personal loss, or the cracks are in a conspicuous position. Often difficult to date first appearance of cracks for this reason. May be associated with wrinkling or wallpaper but easily masked by an embossed or heavy pattern. Typically found at ends

of lintels, junctions between walls and ceilings and between dry lining or ceiling boards, also at junctions of differing materials, for example, where wall-plate at eaves is below ceiling line in upper rooms, and over doors in slender brick or block partitions, whether or not the door framing continues up to ceiling level

| P3 | 1–2 | ▶\|◀ | Cracks more easily observed when approaching 2 mm in width. At this size, cracks can usually be detected by touch as wallpaper is likely to stretch, wrinkle, or tear. In clear light, may be seen at a distance of a few metres. Thus may be detected by careful visual inspection from ground level when investigating upper elevations of two and possibly three storey buildings. May fracture building components such as bricks, blocks, cills or lintels |

| P4 | 2–5 | ▶\| ◀ | Cracks may be conspicuous on most structures, and at upper limit very conspicuous whether the building is well maintained or not. Usually causes property owners to become alarmed, especially if they develop within a few months. At upper limit, draughts may occur through outer walls; doors and windows may 'stick' or fail to close; brick arches may loosen. Likely to be associated with a pattern of cracking in which a mechanism of movement can be discerned originating at support level. Diagonal cracks may develop in ceilings. |

| P5 | 5–15 | ▶\| \|◀ | Cracks of this size are associated with severe damage, especially in the upper range of this class. Doors and windows may jam, glass may crack or shatter, walls crack right through, draughts develop, severe shear patterns may develop including diagonal cracking in ceilings. Cracks may split into two or more parallel fractures leading to shattering of a panel of brickwork. At the upper limit, masonry arches may fail, service pipes |

distort or fracture, noticeable gaps will appear in expansion joints, roof tiling or similar finishes that would normally accommodate or mask smaller movements. Weather-tightness is likely to be impaired. Services may be damaged. Likely to be associated with a mechanism of movement clearly following from differential movement of supports. Support is usually the foundation, but could be sagging floor or beam

P6   15–25   ▶   ◀   Cracks likely to develop in groups with a clear pattern in which a mechanism of movement can be recognised. In older structures there is an increased risk of falling plaster and masonry. Distortion obvious to the naked eye, even at a distance, walls may bulge, especially in older properties. Horizontal movements at bearings and/or at d.p.c. level often develop. Unless the damage is caused by sudden and severe removal of support (for example, mining subsidence, swallow holes, sewer or trench collapse) cracks of this size may have been partially filled in during previous 'making good' operations, thus masking the total distortion

P7   Greater than 25   As for Classes P5 and P6 but damage greater and usually more widespread. The risk of the structure becoming dangerous increases rapidly with advancing maximum crack size, and is greater in older brick or stone masonry buildings. Bearings may be dangerously weakened

If the information scheduled is now considered in the context of a domestic building, a further tabulation is possible and is shown below:

Class of crack	Degree of damage	Effect on structure and building use
P0	Insignificant	None
P1	Very slight	None
P2	Slight ⎫	Aesthetic only, accelerated weathering to external surfaces
P3	Slight to moderate ⎭	
P4	Moderate ⎫	The serviceability of the building will be affected and, towards the upper boundary, stability may also be at risk
P5	Moderate to severe ⎬	
P6	Severe to very severe ⎭	
P7	Very severe to dangerous	Increasing risk of structure becoming dangerous

It will be obvious that the 'Degree of damage' column must be regarded as being in general terms and in this way reflects particularly the traditional two-storey domestic construction of conventional planning. It is important to have in mind always a construction that for some reason is not typical or conditions which are not normal. It will be seen too that, what may be considered severe or dangerous in one construction has a lower rating in another. It is important not to rely totally on classifications or categories in any case and to treat every new situation as special.

Having decided into which classification the cracking should fall, it is important to decide if any temporary work is required and secondly if any permanent work is to be recommended. If the brief is explicit on these points, or if consultants are to be briefed, a whole range of action is possible. It is convenient to consider the two categories of remedial work used by John Pryke, which are detailed in the following list.

Temporary work

(a) Wedging horizontal cracks.
(b) Temporary strutting to bearings.
(c) Strutting openings.
(d) Removal of loose plaster.
(e) Shoring bulged walls.

Permanent work

(a) Routine maintenance.
(b) Chemical rebonding.
(c) Underpinning.
(d) Jacking to lighten horizontal cracks.

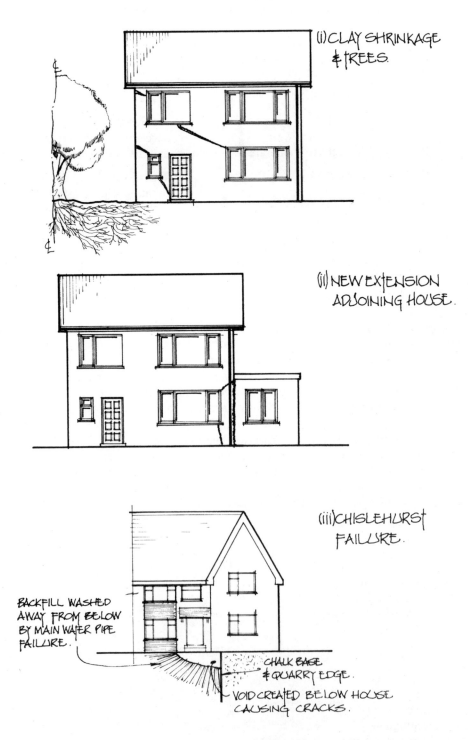

(i) CLAY SHRINKAGE & TREES.

(ii) NEW EXTENSION ADJOINING HOUSE.

(iii) CHISLEHURST FAILURE.

BACKFILL WASHED AWAY FROM BELOW BY MAIN WATER PIPE FAILURE.

CHALK BASE & QUARRY EDGE.

VOID CREATED BELOW HOUSE CAUSING CRACKS.

Figure 14 Cracking patterns

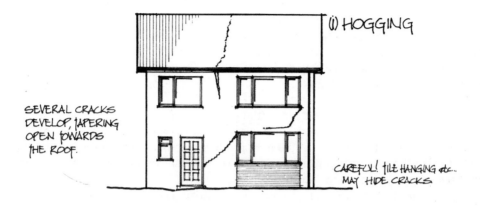

(i) HOGGING

SEVERAL CRACKS DEVELOP, TAPERING OPEN TOWARDS THE ROOF.

CAREFUL! TILE HANGING etc. MAY HIDE CRACKS

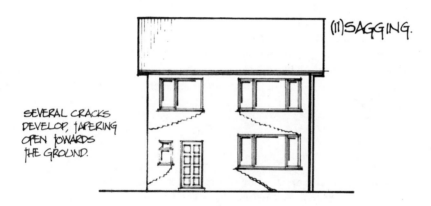

(ii) SAGGING.

SEVERAL CRACKS DEVELOP, TAPERING OPEN TOWARDS THE GROUND.

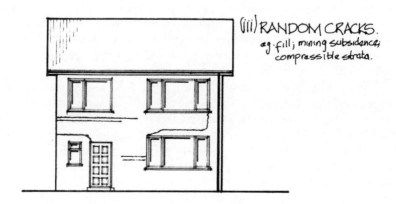

(iii) RANDOM CRACKS.

e.g. fill; mining subsidence; compressible strata.

Figure 15 Cracking patterns

(e)   Easing doors and windows.
(f)   Re-hanging doors and windows.
(g)   Tie-bars.
(h)   Jacking to re-level.
(i)   Re-levelling floors.
(j)   Stitching masonry.
(k)   Installation of jackable foundations.
(l)   Demolition and rebuilding.

If the walls being inspected are 'traditional' then the construction is likely to be cavity walling of brickwork and blockwork with a strong mortar, whether of cement or cement/lime mix. This mortar would have produced strength in compression and shear, but nothing like the same standard in tension. It is important to appreciate the performance of different mortars over time, because if their strengths reduce, and this is common in old buildings, then the pattern of cracking can be modified. For example, mortars initially weak, such as lime mortars, will accommodate some movement, which may be differential and continuing, and therefore such movement may not be observed in a casual visual inspection. Some cracking patterns are shown in figures 14 and 15, which may act as a guide when observing and reporting fractures and when determining likely causes.

The surveyor, when investigating cracked structures, should compare John Pryke's definition scale of cracking with that contained in *BRE Digest 251* 'Assessment of damage in low-rise buildings'. This digest also contains useful advice on the incidence and causes of damage in low-rise buildings and, as such, should be essential reading for all building surveyors.

Wall ties

In cavity masonry walls, the life expectancy of the zinc coating to the steel wall ties appears in practice to be less than expected. In some cases the protection to the steel is well under the notional 60 years' life expected of the walls. The result of this deterioration of the tie can mean problems of stability, not only in old buildings but also in those constructed more recently. During the early stages, rusting ties are not likely to cause any visible damage and so no warning of this defect is usual. However as the situation worsens, an indication of the extent of the corrosion may be horizontal cracking in the joints at normal tie spaces. In many cases, the reduction or total loss of structural connection between the two leaves may occur unnoticed until collapse suddenly occurs.

The seriousness of this condition will depend on several factors, for example, the nature of the loads and the way the walls were originally designed and incorporated in the building. The use of aggressive mortars such as black ash mortar and the use of inferior coatings (that is, bitumen) or substandard thicknesses of galvanising, have also contributed heavily to wall tie failure. The nature of the problem of wall tie failure is outlined in *BRE Information Paper IP 28 / 79* 'Corrosion of steel wall ties: recognition, assessment and appropriate action', and *BRE Paper IP 4 / 81* 'The performance of cavity wall ties'. A performance specification for wall ties has also been produced by the BRE and this is outlined in *BRE Paper IP 4 / 84* 'Performance specifications for wall ties'.

The remedial work could involve on-site replacement with suitable ties, local stabilisation and, in severe conditions, the demolition and re-building of the wall, in part or totally.

A variety of proprietary replacement systems has been developed, including a system of injecting foam into the cavity to stick the two walls together. Choice of repairing system will depend on the construction, the extent of decay and the finance available. *BRE Digest 257* 'Installation of wall ties in existing construction' provides useful advice on the subject. If the building is of the conventional domestic type with a typical plan, some reduction in the strength of the walls due to loss of tie efficiency can be tolerated. This condition could be a more serious consideration if the building is framed and multi-storey with brick in-fill panelling.

Corrosion of steel

In *BRE Digest 263* 'Part 1 Mechanism of protection and corrosion', *Digest 264* 'Part 2 Diagnosis and assessment of corrosion-cracked concrete' and *Digest 265* 'Part 3 The repair of reinforced concrete', attention is given to the durability of steel in concrete. The original protection provided to the steel can be damaged in a number of ways, and when the steel is exposed to air and moisture, corrosion quickly develops. The degree and conditions of exposure will influence the rate of deterioration and the classification of this is

(a) mild
(b) moderate
(c) severe
(d) very severe

The causes of the damage will usually be obvious and in *Digest 264* a flow chart showing the investigation process is reproduced. This is worthwhile reading if a surveyor wishes to assess the structural implications of the defects he has observed. The surveyor can check the construction such as the concrete cover to reinforcement by using a covermeter, but if the building contains prestressed concrete, then the investigation and analysis should be carried out by a structural engineer.

Cavity insulation

In recent years many cavity walls have been upgraded by cavity filling to increase thermal insulation. Difficulties have arisen when this work has been carried out badly or where the existing wall is of poor construction. Clearly if defective workmanship has occurred, then cavity filling will certainly exacerbate the situation. For example, if fissures or voids have occurred in the fill, then water may pass easily from a saturated outer leaf to the finish on the internal leaf. Another point which has been raised in *BRE Digest 236* 'Cavity insulation' is the effect of placing thermal insulation around electric cables that run through the original voids. In these circumstances, overheating and extra loads on the cable can occur.

In the UK, urea–formaldehyde in the form of foam has been used frequently for insulating traditional cavity walls. Generally, it has proved to be an efficient and effective insulant, but in the past the injection has occasionally been performed carelessly or in circumstances that are inappropriate. From these cases there have been a relatively small number of complaints from occupants who have reported discomfort and even irritation of the hair and eyes. *BS 5617* and *BS 5618* were published to specify the quality of the foam

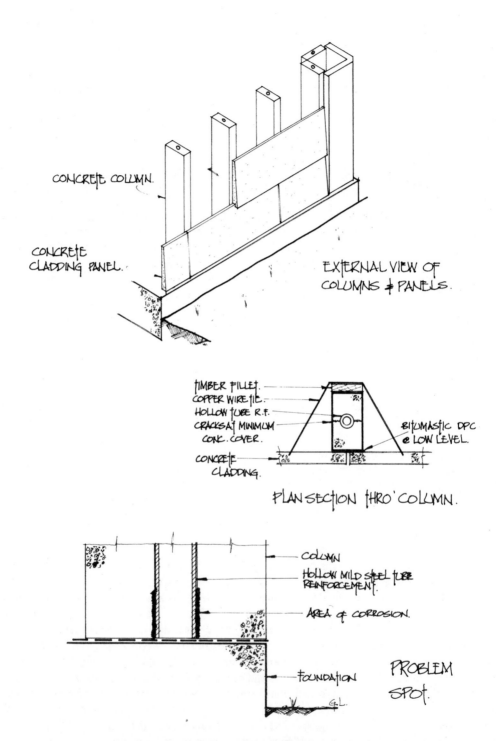

CONCRETE COLUMN.

CONCRETE CLADDING PANEL.

EXTERNAL VIEW OF COLUMNS & PANELS.

TIMBER FILLET.
COPPER WIRE TIE.
HOLLOW TUBE R.F.
CRACKS AT MINIMUM CONC. COVER.

BITUMASTIC DPC @ LOW LEVEL.

CONCRETE CLADDING.

PLAN SECTION THRO' COLUMN.

COLUMN
HOLLOW MILD STEEL TUBE REINFORCEMENT.
AREA OF CORROSION.

FOUNDATION

PROBLEM SPOT.

G.L.

Figure 16 Construction and defect of Airey houses

and the latter is the code of practice for the thermal insulation of cavity walls. If these standards are followed, there should be no risk from the formaldehyde vapour which is inevitably produced during the curing process. Certain constructions are inappropriate for filling by urea–formaldehyde, for example, timber-framed construction and other forms of wall without a masonry inner skin. In these cases where there is a risk of vapour penetration, this form of insulant is not recommended. *BS 8208* gives guidance on the suitability of cavity walls for filling with insulant.

Non-traditional walls

Non-traditional construction was developed shortly after the Second World War and was due mainly to the shortages of skilled labour, the high erection costs for traditional build-ings and the massive demand from Local Authorities wishing to complete their housing programmes. Low-rise, low-cost, steel-and-concrete system buildings were erected in their thousands. The majority of these systems used concrete as their main load-bearing material. These PRC houses fall into two distinct types:

(a) The post and panel dwelling, in which the precast concrete panels span between load-bearing precast columns.
(b) The panel type, where the main load-bearing elements are storey-height precast panels.

Many defects with these early PRC houses have come to light over the last few years, with the main problem being one of corrosion of the reinforcement. This is due in many cases to a lack of sufficient cover to the steel, the carbonation of the concrete and finally the presence of chlorides in the concrete which accelerates the corrosion.

When the surveyor is faced with a non-traditional (PRC) house it may be difficult to determine the type of construction without carrying out extensive investigations. A Department of the Environment publication, called *Housing Defects Act 1984, The Housing Defects (Prefabricated Reinforced Concrete Dwellings) (England and Wales) Designations 1984, Supplementary Information*, contains a number of photographs of the most common types of non-traditional dwellings. Once the type has been determined the surveyor can then identify particular defects by referring to the various BRE Information Papers which summarise BRE reports on the structural condition of prefabricated dwellings. An example is *BRE Information Paper IP 16 / 83* 'The structural condition of some prefabricated reinforced concrete houses of Boot, Cornish Unit, Orlit, Unity, Wates and Woolaway construction' which summarises a number of BRE structural condition reports. One of the most publicised PRC systems which has failed in large numbers is the Airey house, designed by Leeds' builder the late Sir Edwin Airey. Figure 16(i) indicates the typical post and panel construction of an Airey house. Figure 16(ii) shows a section through an Airey precast column, indicating the position and extent of cracking that may be found with this type of construction. Figure 16(iii) shows the position of maximum corrosion which should be investigated during a survey of such houses. A number of repair systems have been devel-oped for the Airey house and other PRC low-rise dwellings, and the surveyor must be able to advise his client on the feasibility of repair and selection of the most suitable system.

6 Roofs

The roof can be considered as the building's primary defence against the weather. If it is breached, decay of other elements, such as walls and floors, will quickly follow. The roof in addition to excluding rain and wind must have adequate fire resistance, thermal and sound insulation and strength and stability. In order for the roof to perform its functions it must be properly constructed with sound materials by competent craftsmen. The main features of a roof that concern surveyors are its shape and form, its construction, the nature of the roof covering and the rainwater disposal arrangement.

To ensure that nothing is overlooked when inspecting roofs, it is advisable to follow a rigid sequence, perhaps commencing externally with the roof covering, ridges and hips, copings, flashings, chimneys, rainwater disposal arrangements, balconies, porches and projecting features, and then internally, the roof construction and insulation. In the course of a reconnaissance survey the surveyor may have to be content with examination of external features from the ground, when a good pair of binoculars is useful. Every effort should be made to see all roof slopes, as well as ridges and hips, and the chimney stacks, but in built-up areas this may not always be practicable because of the impossibility of getting sufficiently far away from the building.

In addition to taking full notes, it is a golden rule to record any feature that it is impracticable to see, and such omissions must be specifically enumerated in the survey report. This is also true if it has not been possible to fill in missing details when inspecting the roof void and rooms on the top floor. Moreover, it should not be regarded as sufficient merely to record what could not be investigated for various practical reasons. If missing information may have a bearing on the surveyor's ultimate conclusions, supplementary instructions should be sought. This should clarify how far the client requires the surveyor to go in calling in builders to attend on him with ladders and other plant, and in negotiating with the vendor as regards further opening up that may cause damage which the prospective purchaser will have to pay for making good.

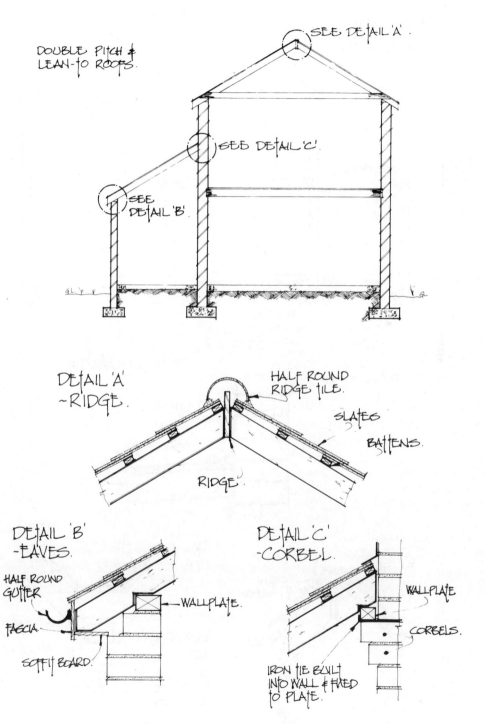

DOUBLE PITCH &
LEAN-TO ROOFS.

SEE DETAIL 'A'.

SEE DETAIL 'C'.

SEE
DETAIL 'B'.

DETAIL 'A'
~RIDGE.

HALF ROUND
RIDGE TILE.

SLATES

BATTENS.

RIDGE.

DETAIL 'B'
-EAVES.

HALF ROUND
GUTTER

FASCIA.

SOFFIT BOARD.

WALLPLATE.

DETAIL 'C'
-CORBEL.

WALLPLATE

CORBELS.

IRON TIE BUILT
INTO WALL & FIXED
TO PLATE.

Figure 17 Simple roof form

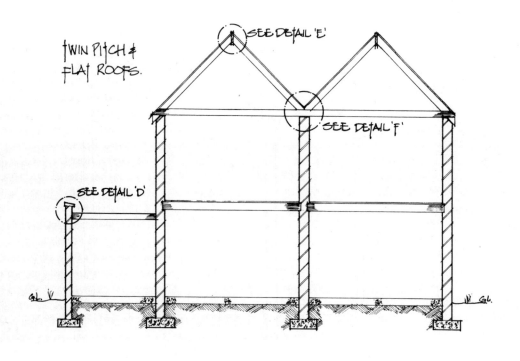

TWIN PITCH &
FLAT ROOFS.

SEE DETAIL 'E'

SEE DETAIL 'F'

SEE DETAIL 'D'

GL.   GL.

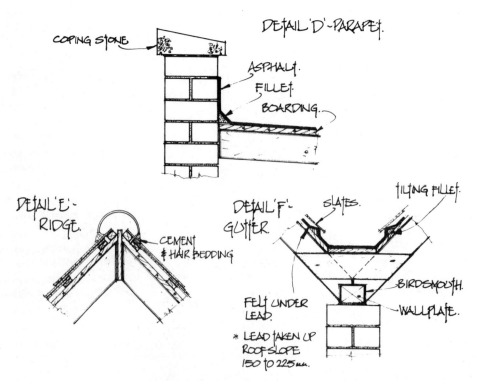

DETAIL 'D'-PARAPET.

COPING STONE

ASPHALT.
FILLET.
BOARDING.

DETAIL 'E'-
RIDGE.

CEMENT
& HAIR BEDDING

DETAIL 'F'-
GUTTER.

SLATES.

TILTING FILLET.

BIRDSMOUTH.

WALLPLATE.

FELT UNDER
LEAD.

* LEAD TAKEN UP
ROOF SLOPE
150 to 225 mm.

Figure 18 Twin pitch and flat roof forms

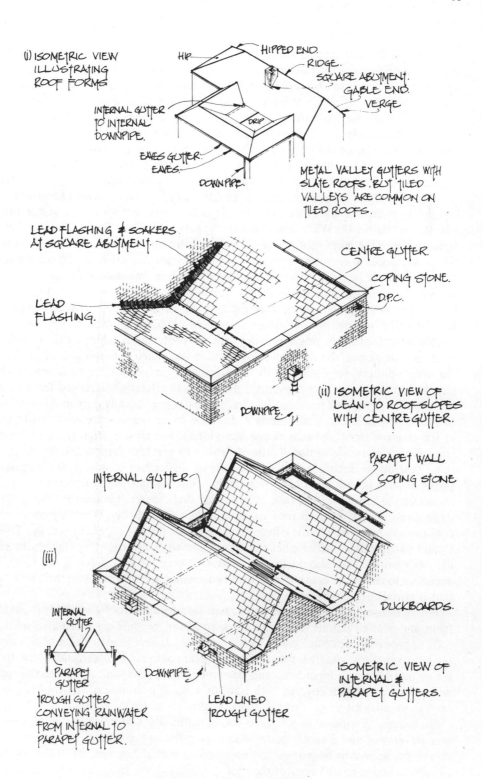

(i) ISOMETRIC VIEW ILLUSTRATING ROOF FORMS

HIP
HIPPED END.
RIDGE.
SQUARE ABUTMENT.
GABLE END.
VERGE

INTERNAL GUTTER TO INTERNAL DOWNPIPE.

DRIP

EAVES GUTTER.
EAVES.
DOWNPIPE.

METAL VALLEY GUTTERS WITH SLATE ROOFS. BUT TILED VALLEYS ARE COMMON ON TILED ROOFS.

LEAD FLASHING & SOAKERS AT SQUARE ABUTMENT.

CENTRE GUTTER.
COPING STONE.
D.P.C.

LEAD FLASHING.

DOWNPIPE.

(ii) ISOMETRIC VIEW OF LEAN-TO ROOF SLOPES WITH CENTRE GUTTER.

PARAPET WALL
COPING STONE

INTERNAL GUTTER

(iii)

DUCKBOARDS.

INTERNAL GUTTER
PARAPET GUTTER

TROUGH GUTTER CONVEYING RAINWATER FROM INTERNAL TO PARAPET GUTTER.

DOWNPIPE

LEAD LINED TROUGH GUTTER

ISOMETRIC VIEW OF INTERNAL & PARAPET GUTTERS.

Figure 19   Internal gutters

Roof forms

The simplest form of sloping roof is the mono-pitch or lean-to roof, but the most common form is the double-pitch, the two slopes meeting at the apex in a ridge as shown in figure 17. Such roofs are used both for cottages and large buildings, with spans up to 12 metres (40 ft). They may terminate at both ends in gables, the pitch roof over-sailing the end walls of the building, the last rafters at each end being protected by a fascia board or barge board. Alternatively, the end walls may rise above the roof slopes where they are capped with coping stones or the slopes may be hipped back at both ends.

For spans of more than 12 metres (40 ft) the double-pitch roof is normally uneconomic because of the excessive height from eaves to ridge, and the complexity of the construction. Therefore, in older domestic work, the problem of spans larger than 12 metres (40 ft) were frequently overcome by using twin double-pitch roofs, with a centre gutter between the two slopes, figure 18. With other than rectangular plans, subsidiary double-pitch roofs at right-angles to the main roof offered a simpler solution, except for the necessity for valleys where two slopes at right-angles meet. The older, larger country house, square in plan, often has a completely internal gutter surrounded by the four internal roof slopes, figure 19(i), or the area between the four external roof slopes may have a slightly domed flat roof.

In urban development in terraces eighty years or more ago, architects appear to have been at pains to hide the roof slopes from view. A common solution was to provide twin double pitch roofs with ridges parallel with the roof, and a centre gutter between the two internal roof slopes, the rainwater from which was taken through one or both roof voids via trough gutters, see figure 19(iii). By erecting a parapet wall or pediment, enclosing the parapet gutter on the front elevation, the roof was effectively screened from view. Access to the centre gutter was provided by means of a dormer, usually only in one roof slope, and then often far from easy to negotiate. Ladders or cat walks were provided to give access to the external front and rear slopes and gutters, but all too often these are in a decayed state and quite unsafe to use. Trails of broken or cracked slates mark the routes taken by those sent up to clean out gutters. Because of difficulties of access, the gutters and roof slopes of such houses are all too frequently neglected until water penetrates the ceilings of rooms on the top floor. The task of surveying such houses is made the more difficult since there is rarely entry to both roof voids, and the accessible one is usually obstructed by the cold storage tank, which it is often virtually impossible to inspect internally. The surveyor should particularly note whether the access hatch is large enough to allow the renewal of the old storage tank, should the need arise. Where the access hatch is small, it is often necessary to renew the existing tank with two smaller ones coupled together, an arrangement that is not always satisfactory.

Where houses are of no great depth, two lean-to pitch roofs with ridges parallel to the front and rear elevations, and with a centre gutter, provide an alternative solution to the roofing problem, the centre gutter sometimes being almost a small flat roof. Rainwater from the centre gutter is usually taken through the rear roof void into a trough gutter to discharge via an outlet into a hopper head and downpipe on the rear wall. Access to the centre gutter will consist of a hatch covered with lead, or a skylight, in one of the roof slopes, which is even more difficult to get through than the dormer of larger houses.

On narrow frontages two lean-to roofs, with ridges at right-angles to the front and rear elevations, and a centre gutter discharging through an outlet into a hopper head and downpipe, appear to be particularly common in central London, see figure 19(ii). There is usually an access hatch in one of the roof slopes, again far from easy to negotiate. Although the centre gutter is usually very narrow to walk along, it is to be preferred for obvious reasons to a trough gutter taken internally through a roof void. In country districts there is often no access hatch or dormer to a centre gutter between two internal roof slopes, the only means of access being either an attic window looking on to the centre gutter, or a ladder from the ground. The condition of the centre gutter and internal roof slopes is so important that the surveyor must acquire ladders of adequate length to permit inspection.

Rooms are often formed in the roof spaces of pitch roofs, involving dormer windows or mansard slopes. Every effort must be made to check that the openings for windows have been properly constructed, that the rafters performing the function of trimmers are of larger section than rafters between the openings, and that the dormer cheeks and roofs are sound. In a reconnaissance survey the surveyor will be restricted in the extent to which he can check on the construction of the dormers if there is no access door in the ashlaring. In a detailed structural survey it is imperative to have holes cut in the ashlaring to permit a thorough examination, not only of the dormer construction, but also of the feet of the rafters, wall plates, and the ends of the floor joists of the attic floor. Moreover, the condition of the plaster laths of the ashlar walling will also warrant investigation. The significance of any beetle infestation present in the laths must be assessed.

Externally, the condition of dormer cheeks and roofs is very important. If there are parapet gutters in front of dormer windows it is usually possible to get out into the gutter, when they can be adequately inspected. Failing such access, it may be possible to see the cheeks and roof by sitting on the window board, or base of the window frame, getting out backwards as far as is safe. If the dormer windows are double hung sashes there is likely to be insufficient room for the surveyor to get out, when he will either have to gauge the condition of the dormers from the state of the plaster on the internal faces of the cheeks and soffit or the dormer roof, or defer completion of his survey until ladders are available. Since dormers of old properties are prone to decay, every effort must be made to examine them adequately. The bases of dormer frames and jambs must be probed and any evidence of water penetration noted. In older properties that have not been exceptionally well maintained the plaster on the cheeks and soffit is likely to be perished, and the external sills and bases of frames and jambs are likely to contain wet rot. Externally, the cheeks and roofs may have been covered with zinc, and nothing short of complete reconstruction of the dormer opening and window frame can be recommended. Since the reconstruction of a dormer window, at no great height from the ground, may well cost £1000 or more, it is imperative to give them the closest attention.

Roof coverings

In the UK a variety of roof coverings abound which, before mechanised transport, reflected the character of the surrounding area. Except for a few important public buildings, which may have roof slopes covered with metal such as copper or occasionally lead, traditional covering materials are slates and tiles, and where stone is extensively used for walls, stone slabs or stone slates will often be found. In rural areas, thatch and sometimes shingles were used on dwellings while asbestos–cement slates or tiles and corrugated steel sheeting were usually confined to the roofs of outbuildings or substandard properties.

Corrugated steel sheeting, used as temporary war damage first aid, may still exist on hidden slopes of otherwise good-quality properties. When examining roof coverings it is essential to check the abutment of such coverings with other elements. Any change in direction or perforation through the roof covering will warrant closer inspection as will chimney stacks and flashings because of their exposure.

Thatch

Thatch provides a well-insulated roof covering, with a life varying from up to twenty years for straw thatch and up to sixty years for reed thatch. This roof covering is only suitable in rural areas and then, preferably, only for detached properties, well away from any neighbours. Even so, it provides a fire hazard which is reflected in high insurance premium rates, both for the building and its contents. The problem of the high cost of insurance should certainly be stressed in the survey report, because it may come as a shock to the prospective purchaser.

From the surveyor's standpoint the condition of the thatch is obviously important, but the roof construction no less so. In many period cottages and houses where thatch has been used, the rafters are often found to be no more than half-round 'hardwood' poles of whatever species was readily obtainable in the district. As often as not, these half-round poles contained upwards of 75 per cent of sapwood. If a susceptible timber, it is likely therefore to have been attacked, first by *Lyctus* infestation and subsequently by the common furniture beetle. If the attack is severe, only a fraction of the original sectional area of the rafters may be of any structural value. *In-situ* insecticidal treatments are of no practical value, even where there is continuing active beetle population (unlikely in an old cottage or house), because the chemical treatment does not restore the strength properties of the attacked timbers. Moreover, the most common ties used for tying bundles or reeds and straw together before and during laying are of hazel, which is particularly susceptible to *Lyctus* infestation. Hence, although the condition of the thatch may suggest that it has a further serviceable life of several years, the condition of the hazel ties and half-round poles used as rafters may necessitate advising complete reconstruction of the roof and renewal of the roof covering, at an earlier date. Such adverse comments will invariably disappoint the client, who has probably allowed the picturesque appearance of the property to overrule all practical considerations.

Wood shingles

Wooden shingles are likely to be found only on comparatively modern buildings and will almost certainly be of rift-sawn western red cedar. This is a particularly decay-resistant timber, and shingles laid to an adequate pitch (that is, not less than 45°), are likely to outlast the nails with which they are fixed. The Canadian authorities, who have extensive experience in this covering, recommend aluminium nails or, as a second choice, hot-dipped galvanized nails. The proximity of trees to shingle roofs is important because the constant dripping from overhanging branches has a particularly adverse effect on their life.

Slate and stone roof coverings

Slates are quarried in several parts of the UK, but production has dropped off considerably in the last few decades since man-made asbestos composition slates have become available.

Slates have also been imported from continental quarries some of which are as good as home produced slates, but many of them are of inferior quality.

The slates from North Wales are mostly blue, blue-grey, and blue-purple; from South Wales they are green and silver-grey; from Cornwall grey, green, and russet-red; from North Lancashire grey-blue; from Westmorland various shades of green; and from western Scotland bluish-black and black. After quarrying, the individual slates are produced by splitting the quarried blocks along natural planes of cleavage.

The slates from North Wales are available in several different sizes, the whole of one roof slope being laid in slates of the same size, and of much the same colour. Slates from South Wales, Westmorland and Western Scotland are in random sizes, and are laid in graduated diminishing courses from the eaves to the ridge. The sorting to size and varying the spacing of battens up the roof slopes can add appreciably to the labour cost, compared with laying those of uniform sizes. The slates from Cornwall and North Lancashire are available both as random and sized slates.

Westmorland slates are appreciated for their decorative qualities, but they contain a high proportion of calcium carbonate. In spite of this, the slates have a very high reputation for lasting qualities, both in London and other large cities. The slates from South Wales, Cornwall, North Lancashire, and the bluish-black (but not the black slates) from Western Scotland also have a reputation for durability. The sized blue slates from North Wales are very widely used, and they make a satisfactory, relatively light-weight roof covering, which is less expensive to lay than random-sized slates. They do, however, lack the aesthetic qualities of the more variable slates from other districts. The different roofing slates outlined previously, vary appreciably in weight from as little as 27 kg/sq. metre ($5^1/_2$ lb/sq. ft), for the best Bangor slates up to 49 kg/sq. metre (10 lb/sq. ft), for Thirds, compared with 44 kg/sq. metre (9 lb/sq. ft), for Best Westmorland up to 73 kg/sq. metre (15 lb/sq. ft), for Thirds. Stone slates and slabs are appreciably heavier than slates, being about 88 kg/sq. metre (18 lb/sq. ft), to little short of 146 kg/sq. metre (30 lb/sq. ft).

Concrete 'slates' to imitate stone are now being produced. These concrete slates are a reasonable imitation of the genuine product but, being regular in shape and size, they lack the interest of a true stone slate roof which has been laid in diminishing courses. In addition, they do not weather to the attractive shades of natural stone and, as yet, there is no reliable data for comparing the life of concrete slates with that of natural stone.

It will be appreciated that more substantial battens and roof scantlings are called for with the heavier stone roof coverings. The fault of such roof construction is usually the lack of adequate framing together of the different roof members, rather than the inadequate size of scantlings. In ancient buildings the roof construction is almost invariably less substantial than it might otherwise be because rafters and purlins were usually fixed with the lesser scantling dimension in depth, in other words, showing the wider face to the stone covering. Other defects of such old roofs are likely to be associated with subsequent beetle infestation or fungal decay, but poor selection of timber is also a factor. In addition, the proportion of sapwood in different members is likely to be very variable and large knots that even the worst 'jerry builder' would reject today, were accepted by the real craftsmen who framed up these old roofs with such skill and care.

The condition of the roof covering viewed from above is often misleading, and the surveyor should endeavour to check on the condition of the slates or stone slabs from inside the roof void. It is only from inside the roof that the means of fixing can be established. Roof coverings of buildings 100 years of age or more are likely to be fixed with wooden pegs, and the surveyor should know sufficient about timber identification to be able to determine whether the pegs are of oak or a softwood. If they are softwood, large numbers of pegs are likely to be decayed, necessitating early stripping and recovering of the roof. Oak pegs on the other hand, may still be remarkably sound, even outlasting the roof covering itself. The danger with wooden pegs is the likelihood that when pegs failed because of fungal decay, the slates or stone slabs were re-fixed by bedding in mortar. This makes it very difficult to get at adjoining slates or slabs when carrying out repairs, without breaking those bedded in mortar. Iron nails succeeded wooden pegs and, when of sufficient gauge, they had a serviceable life of forty to sixty years. Thin wire nails on the other hand are very inferior, and are unlikely to have a life of more than thirty years. In high-class property, copper or yellow composition nails (an alloy of copper, zinc and tin) were often used, and these are very satisfactory and likely to outlast the roof covering. More recently, galvanised nails have been widely accepted and, provided they are coated with an adequate thickness of zinc, they can be expected to give a reasonable service life. Still more recently, aluminium nails have become available and, if of sufficient gauge and length, and with shanks conforming to the recommendations in *BS 1202: Part 3: 1974*, then they will be quite satisfactory.

Tiles

Tiles differ from true slates, stone slates and stone slabs in that they are a manufactured product, made either from natural clay deposits (clay tiles), or from cement, colouring agents and sand (concrete tiles). There are two classes of both clay and concrete tiles, the first being the plain tile which is a comparatively flat unit and the second being the single-lap interlocking tile whose distinctive feature is its corrugated profile. Plain tiles are laid to laps varying from 65 to 100 mm ($2^{1}/_{2}$ to 4 inches) and single-lap interlocking tiles, including Pantiles, Double Roman and many other varieties, are laid to differing laps, depending on their profile. Tiles may be classified in four categories: hand-made, machine-made, machine-made sand-faced, and concrete. At one time all tiles were hand-made exclusively from local clay deposits, and once extracted they were compressed to the size and thickness required, allowed to dry, and then fired in a kiln. Today, machine-made clay tiles are identical in composition to the hand-made product in that they are shaped from natural clay deposits, dried, and then fired, but with improved methods of manufacture starting with the preparation of the clay through to drying and firing. The best machine-made clay tiles of today possess better characteristics for longevity, homogeneity of the clay and freedom from lamination. Plain tiles will have a very long life if laid to a sufficient pitch (that is, not less than 40°), however, in older buildings, slopes may have to be stripped and re-covered, reusing the bulk of the original tiles, because of failure of the wooden pegs or nails with which they were originally fixed. On older buildings where the tiles appear irregular in colour, shape and size, then it is likely that they will be hand-made. There is a texture, mellowness and absence of rigid conformity in hand-made clay tiles that cannot be matched by the machine-made product, although the tile of today is a very great improvement on the shiny tiles produced at the turn of the century.

In the past, single-lap interlocking tiles, including the various types of pantiles, were less uniform and often exhibited a wide range of sizes and profiles, even those from the same batch. These generally were of foreign manufacture, particularly from the Netherlands. Such tiles can be laid to a pitch as low as 30°, but the roof slopes should have been boarded, battened and counter-battened and felted with a breather-type paper. Some of these single-lap tiles were nailed through the head, some through the nib, and others have special clips at the sides or bottom of the tile, or they were wired through the lug on the underside.

Today, tiles used on new housing estates and the smaller private houses are almost invariably made of concrete and usually of the single-lap interlocking type rather than the plain tile type. With plain tiled roofs where tiles are only holed for nailing, each tile has to be secured with nails, although formerly these were likely to be wooden pegs. Tiles that are both holed and nibbed should normally be nailed at every fifth course, relying on the nib at the top of the tile clipping the tiling batten and the weight of the overlapping tile to hold down those not nailed in position. Nails used for fixing tiles are similar to those used for fixing slates, but of appropriate gauge. In good-quality work the nails should be of copper or composition alloy, but the majority of tiles are likely to be fixed with galvanised nails and more recently, aluminium.

Hand-made plain clay tiles often vary in size from 250 x 150 mm (10 x 6 inches) for tiles holed for nailing, to 275 x 175 mm (11 x 7 inches) for tiles holed and nibbed. The standard size for holed and nibbed machine-made clay or concrete plain tiles is 265 x 165 mm (10½ x 6½ inches). Weights will vary but for double-lap hand-made clay tiles laid at a 40 pitch, the installed weight is approximately 78 kg/sq. metre (16 lb/sq. ft.), and for double-lap machine-made, plain clay tiles, 68 kg/sq. metre (14 lb/sq. ft). Concrete plain double-lap tiles laid at the minimum pitch of 35° weigh approximately 78 kg/sq. metre (16 lb/sq. ft).

Interlocking single-lap clay and concrete tiles are larger than plain tiles and are produced in a much greater variety of sizes ranging from 290 x 215 mm (11½ x 8½ inches) to 381 x 225 mm (15¼ x 9 inches). Weights also vary between 39 kg/sq. metre (8 lb/sq. ft.) and 50 kg/sq. metre (10 lb/sq. ft.). The interlocking tiles can be laid to pitches as low as 18° although the average is around 22°.

Manufactured tiles should be made in accordance with the recommendations in *BS 402: 1979* 'Clay plain roof tiles and fittings', and *BS 473 and 550: 1971.* 'Concrete roofing tiles and fittings'. Various special shapes of tiles are made for forming ridges, hips and valleys.

Ridges and hips

Apart from lead ridges and hips, which are fixed with nails to timber rolls, the ridge and hip units will be bedded direct on to the roof covering with mortar. Where two units abut one another, the joints are pointed with mortar which should contain a proportion of stone dust, if the ridge and hips are of natural or reconstituted stone. Again, it is a common failing of those who bed and point ridges and hips to use too strong a mix, with the result that shrinkage of the mortar occurs, and the bedding material and pointing falls out. Missing bedding and pointing must be recorded and the importance of making good these defects stressed, because of the risk of rainwater blowing up under the ridge, or penetrating through the joints, creating conditions favourable to fungal attack in the timbers at the apex of the roof. At the ridge the fungal attack is more likely to be wet rot rather than dry rot. This is because temperatures within the roof void in hot summers are likely to be too high for the commonest form of dry rot. With hips it is important that the hip unit is properly bedded particularly if there is no bracket (hip iron) to prevent it from slipping off the roof. The owner would be liable in negligence to any passer-by injured by any part of the roof falling on to him if the cause were lack of maintenance. Similarly, the surveyor owes a duty to his client to observe and report on such defects.

Slate ridges and hips on buildings of some age appear to be particularly prone to splitting, and replacements may be virtually unobtainable, or at any rate, decidedly costly. Defective slate ridges and hips of buildings 100 years old or more are likely to have to be replaced *in toto* because of the impossibility of obtaining replacements to match. Stone ridges may last for a very long time and then disintegrate suddenly under exceptional weather conditions, for example, a few days of severe frost following a prolonged, excessively wet period.

With lead ridges and hips the important considerations (apart from sufficiency of overlap in the separate lengths of lead) are the state of the lead, the condition of the nails securing it to the timber rolls, and the condition of the rolls themselves. The nails are usually broad-headed and either of galvanised iron or copper. It is unlikely that anyone using lead rolls would fix lead with iron nails, liable to rust and corrode quickly, but it is unwise to take anything for granted as unskilled labour used on repairs can make the most extraordinary mistakes because of lack of technical knowledge. There is a tendency for the nail-holes through lead to become enlarged, particularly in hips, and hence the heads of nails through lead should be covered with lead dots. The life of galvanised nails is very variable depending on the thickness of the zinc. Copper nails are satisfactory, but expensive. Composition or yellow metal nails are also suitable, and probably less expensive than copper. More recently aluminium and silicon–bronze nails have been used, the latter only being available to special order, and expensive. The different types of nails are dealt with in *BS 1202: 1974: Parts 1, 2 and 3*.

Copings

Copings should be in a hard, natural or reconstituted stone, or concrete so that they withstand conditions of severe exposure. All too frequently however, the top of a parapet wall or gable end is merely rendered or covered with a course of bricks on edge, laid on creasing tiles as a damp-proof course. Copings may be weathered both sides of a centre line, or laid with a weathered face sloping to one side. There should be throats both sides on the underside of the over-sailing portions of copings. The throat should be a least 12 mm ($^1/_2$ inch) from the outer edge of the coping, semi-circular in section, 12 mm ($^1/_2$ inch) in diameter, and 25 mm (1 inch) out from the vertical face of the wall on which the coping is bedded.

Copings should be bedded on an impervious damp-proof course; ideally there should be a second damp-proof course, just above the gutter or roof line. It is exceptional to find a damp-proof course in either position in older buildings. At the most, there may be two courses of tiles in lime mortar, which do not constitute an impervious barrier to water penetration into the brickwork or masonry below. The soft stone copings of older buildings are all too often in a deplorable condition, spalling on the top surface and breaking away on both edges, rendering them both completely ineffectual as a protective covering for the wall, and positively dangerous to those passing underneath because of slippage or lateral displacement. Well-burned, purpose-made, half-round bricks bedded on a lead-cored damp-proof course, make a very suitable coping for a garden wall or parapet wall of an outbuilding.

The pointing of joints in the separate units forming a coping is also important. All too frequently, too strong a mix is used and it is found that water is penetrating at every joint. In old buildings brass or copper cramps are frequently found linking two coping stones. There is a tendency for the grooves and mortices formed in the stone which hold the cramp to become enlarged when water penetrates both the joint and the coping stone, hastening deterioration. The surveyor should always look for evidence that iron cramps have been used with stone as oxidation of the iron will, in time, burst stones in which the iron is embedded.

Because of neglect of maintenance, the absence of damp-proof courses or the use of unsuitable coping materials, the masonry below the coping and the coping itself will frequently be in a very poor condition. It is not unusual to see vertical fractures in brick parapet walls and in stone parapets backed with brickwork. The trouble is usually sulphate attack which occurs because of the chemical reaction, in the presence of water, between tricalcium aluminate present in ordinary Portland cement and the soluble sulphates within the clay bricks. The chemical reaction results in expansion of the mortar joints and, if the amounts of soluble sulphates in the bricks are high, the bricks themselves may split, or they may disintegrate from the face inwards. The sulphates can only be transferred from the bricks to the joints by percolating water, hence the importance of sound copings, with water-proof joints on parapet walls, free-standing walls and gable ends. The copings should be bedded on a good-quality damp-proof course such as lead-cored bituminous felt. Parapet walls are particularly vulnerable because of their exposed position and, if there are no throats on the underside of the over-sailing coping stones, water will run down the parapet wall, saturating the brickwork for long periods, and creating conditions ideal for sulphate attack. *BRE Digest 89* 'Sulphate attack on brickwork' makes recommendations for minimising the risk of sulphate occurrence, both in facing brickwork, and in parapet and free-standing walls.

Flashings

Flashings are essential where chimney stacks emerge through roof coverings and where sloping roofs abut vertical walls. All too frequently these flashings are only cement fillets, very liable to craze or shrink away from the vertical faces of the stack or wall. It is a common failing to use a mix as strong as 1 part of Portland cement to 3 parts of sand, and such a mix is liable to shrink and craze within a year, making the cement fillet useless for keeping out water. It is important to draw attention to such defective cement fillet flashings because their replacement with proper lead soakers and flashings is expensive. If there are soakers under the cement fillets, these are liable to rapid corrosion unless they have been coated with bitumen prior to fixing.

Where roof slopes meet the vertical faces of stacks and walls, flashings should be in lead, tucked into a joint of a course of brickwork or into a chase in masonry, and then wedged and pointed. It is important to see that such cover flashings are in position, and well pointed. Where loose and away from the stack or wall, the cause may be insufficient tuck-in of the flashings, or the face of the brickwork or masonry may be eroded until no chase is left into which they can be re-dressed and re-pointed. In such circumstances it is likely to be necessary to re-face the brickwork or masonry so that a proper fixing can be provided. Where roof slopes meet vertical faces of stacks, or gable ends and pediments, on the splay, there should be lead soakers, covered by stepped lead flashings. It is often only possible to ascertain whether soakers are present from inside the roof void, when the surveyor should satisfy himself that the soakers are of the same metal as the flashings, otherwise electrolytic action between two different metals may result in rapid corrosion.

Where flashings are found to be in zinc, it is usually an indication that cost has been a paramount consideration and the surveyor should be on the alert for other evidence of want of adequate maintenance.

Gable ends that rise above roof slopes are exposed to the further hazard of defects in the flashings where the roof slopes abut the gable ends. If these flashings are comprised of cement fillets with no soakers beneath, the brickwork or masonry abutting the roof slope can remain in a saturated condition for long periods. This may cause rapid deterioration of the wall and increase the likelihood of decay in any roof timbers in contact with the gable ends.

Chimneys

In old buildings, chimney stacks may be out of plumb without rendering them necessarily dangerous and in need of re-building. Lack of plumbness will be of much greater significance in a tall free-standing one-flue stack, compared with a multiple-flue stack only rising a few feet out of the roof. The state of the pointing must be noted, for in an otherwise well-maintained building the pointing of chimney stacks often leaves much to be desired and, because of the cost of scaffolding, even the simple operation of re-pointing a stack can be costly.

Stacks should be carefully examined for any fractures, or staining by flue gases. Water vapour produced in large quantities in domestic flues can lead to condensation penetrating the chimney walls. Salts and other products of combustion can be carried by this moisture through the wall to appear externally on the chimney stack. This will eventually lead to decay because of the chemical reaction caused by the condensate, which is by sulphuric acid dissolving the mortar. The sulphates produced in this way then attack the Portland cement or hydraulic lime in the mortar, causing it to expand. Expansion may be sufficient to produce fractures in the brickwork or masonry of the stack, and even vertical splits through the brickwork or masonry. The risk is much greater when there is a change over to slow-burning fuels in enclosed grates or boilers after many years of burning coal in an open fireplace. With the passage of time, flue walls will have become coated with hard deposits containing a considerable proportion of chlorides, which cannot be removed however thoroughly the flue is swept. The problem is often encountered when gas-fired boilers are installed in old flues without taking precautions to line the flue adequately. This is because the products of complete combustion of domestic gas supplies are water vapour and carbon dioxide, and the temperatures of these flue gases being relatively low, the resulting condensation accelerates the chemical reaction between the sulphur deposits and the mortar joints. Chimneys are also liable to damage by frost as they are often frozen when saturated with water.

In an extreme case the installation of a gas-fired central heating boiler in a London house split the flue from the first floor upwards in the space of eighteen months, and sufficient water seeped through the fractured brickwork to cause an outbreak of dry rot where the plaster wall finish was applied to battens fixed to the brickwork of the chimney breast.

Rainwater disposal

Whether it is a sloping roof or a flat roof, the surveyor must pay particular attention to the disposal of rainwater. The lead or copper (it is to be hoped that the metal is not zinc) in parapet, centre, internal, trough, and valley gutters should be carefully examined for any splits or repairs. Where the repairs are merely soldered patches these warrant the closest scrutiny because they frequently have a short life. This is particularly so if the sole of the gutter is wide and the parapet wall is not high enough to shade it from the heat of the sun. The position of large splits or old repairs should be noted on a rough sketch drawing so that the condition of the timbers under the splits or repairs can be checked during the inspection of the roof void. In one case, failure to do this resulted in a widespread outbreak of dry rot being overlooked when a mansion was being re-decorated to a very high standard and at great cost. A year later a fruiting body appeared on a door lining and the extent of the attack had to be traced. The plastered wall face of the bedroom underneath was applied to battens and attack had spread down the battens, with the result that not only door linings and lintels were involved, but also much of the wall face and expensive new decorations.

A common failing of older buildings is the length of the bays in parapet gutters which, preferably, should not exceed 2250 mm (7 ft 6 inches) between drips. With the heavier lead used 100 years ago or more the size of bays may not have been so important, but now that lead heavier than Code 5, 25.4 kg/sq. metre (6 lb) is rarely used in gutters, re-laying often involves forming additional drips, which may be difficult to accommodate particularly if the parapet wall is low.

It may not be possible to determine how far up the roof slope the lining has been taken, but every effort should be made to establish this inside the roof space. The amount of the upstand against the parapet wall must be established by turning back the cover flashing. It is to be hoped that the surveyor will not encounter the thoroughly bad practice of taking the upstand up the parapet wall for a short distance, and then tucking it into the wall with no cover flashing; such a case has been met in an expensive property in central London.

If the surveyor has to advise re-laying the gutter, he should always recommend renewing the gutter boarding in pressure-treated timber. By taking up the boarding of a parapet gutter it is possible to check on the condition of the bearers and soldiers and the wall plate, which it may well not be possible to probe or even see adequately from inside the roof void. When parapet and centre gutters have been re-lined relatively recently, zinc, bituminous felt or asphalt may have been used in lieu of lead or copper. As already indicated, zinc is not a suitable material, but felt or asphalt can be satisfactory if properly laid, provided their comparatively short life of say up to thirty years is appreciated. The surveyor then has to consider whether, in changing over to a cheaper material, the previous owner has insisted on the repairs being carried out thoroughly, including searching for other ancillary defects.

Outlets through walls should receive attention, and also the condition of hopper heads into which they discharge. If these hopper heads are in cast iron and tight to the wall, it is more than likely that the backs will be badly corroded, necessitating renewal. The sizes of downpipes are also important and these should have been noted in the course of the external inspection. Size should be governed by the area of roof to be drained and the number of separate outlets and, in consequence, the number of downpipes. In most cases a 75 mm (3 inch) diameter pipe should suffice for an ordinary domestic property, but if the outlet is large, as from a main centre gutter, a 100 mm (4 inch) downpipe may be required. On large public buildings larger diameter pipes are used. Where the roof area is small, as for example a small flat roof over a bay window or porch, downpipes as little as 50 mm (2 inches) in diameter may theoretically be ample, but such small-bore pipes are very liable to become blocked in autumn by leaves.

Those who designed parapet gutters half a century or more ago did not make the mistakes that those who have no experience of the maintenance of old buildings are making today. Where an additional storey has been added, particularly if a mansard roof has been introduced to gain floor area, the parapet gutter may be so reduced in width that it is impossible to repair or replace the gutter lining without taking down the parapet wall. This is a point that the surveyor should always watch for and bring to the attention of the client. The linings to trough gutters are much less likely to call for adverse comment compared with parapet gutters, being protected by the roof covering from the direct rays of the sun. The commonest faults of trough gutters are their shallowness and the inadequacy of the openings at the inlet and outlet ends. The route of a trough gutter should be noted, as this may well lead to discovery of outbreaks of dry rot in lintels or linings of doors openings in the middle of the building. It is equally important to note the line of centre gutters between two internal roof slopes. A common failing in an internal gutter between two pitched roofs of houses on a narrow frontage is settlement in the length of the gutter, caused by shrinkage of the bearers and other timbers under the gutter. Where the fall in both directions to the inlets of the trough gutters was small in the first place, slight settlements can result in the fall being reversed, so that water builds up in the gutter, finding its way up under the roof covering on to ceilings of the rooms underneath. This is where a spirit level becomes essential and it is always advisable to check the amount of fall in an internal gutter.

Eaves gutters can usually only be inspected with the aid of ladders, although the commonest defects of leaks at joints and gutters out of alignment can readily be seen if it rains while the survey is in progress. The alignment of eaves gutters around bay windows and balconies is of particular importance, especially if the stopped ends are tight to the main wall of the building. If the gutters are out of level and fall towards the stopped ends, rainwater is likely to run down the main wall of the building. There will often be the ends of a bressummer in the wall at the back of the stopped ends of such eaves gutters, often no more than 112 mm (4^1/$_2$ inches) behind the face of the wall and, if the gutter is out of alignment, there is a considerable risk of decay developing in the bressummer. Particular attention should be paid to the condition of brickwork under such stopped ends in order to detect tell-tale marks of water running down the wall.

A warning has already been given regarding box gutters on tops of walls where buildings have been built up to the owner's own boundaries. This construction means that the owner cannot clean out such a gutter, replace slates or tiles on the roof slope, or re-point the wall without the permission of the adjoining owner. Moreover, in the course of a survey the surveyor may not be able to inspect the gutter, roof slope or wall if the neighbour will not grant him permission to go on his land. The serious disadvantages of buildings built tight up to the boundary must be stressed in the survey report. Even if the owner has been able to obtain permission from his neighbour to go on to the land to erect ladders and re-paint gutters, it is certain that painters will not have removed the gutter to permit painting of the underside where it is bedded on the wall. In consequence, the sole of the gutter is very liable to become corroded and leak. If this is the case, the wall below will frequently be saturated. Wall-plates and the feet of rafters in such circumstances are very prone to become attacked by dry or wet rot, and their condition warrants the closest attention when continuing the survey of the building.

Balconies, porches and projecting features

It is appropriate after completing the survey of the main roof and the rainwater disposal arrangements to direct one's attention to any flat roofs over projecting porches, bay

windows and any balconies. All too frequently these features are laid to falls towards the main external wall of the building, so that the outlets and downpipes can be tucked out of sight, behind the projecting features. Moreover, the bore of the outlets and of the downpipes is often very small, rendering the outlet liable to blockage. When this happens water will build up on the flat roof, often rising above the skirtings to asphalt roofs or above the cover flashings if the roof is covered with lead or copper. Any timbers such as wall plates, joists or bressummers, housed in the external wall behind such roofs, are then liable to attack by dry rot. If the rooms inside have parquet floors, the surveyor will not be able to expose such timbers in the course of a survey. The most he can do is to probe skirtings and examine the base linings to window openings. In good-class property, skirtings are often as much as 32 mm ($1^{1}/_{4}$ inches) thick, and merely probing the surface may not disclose decay at the back of such skirtings. If there are sills low down or thresholds to 'French windows', then the underside of these should be very thoroughly examined. The surveyor should also look for water staining of the flooring against 'French windows', and for gaps between the base of the skirtings and the floor level of the suspect floors. Walking along the floor close to the external wall may reveal squeaking in the floor. It is also advisable to check the soffits of window and door openings in the room or entrance hall underneath.

It is important to set out very clearly in the survey report the extent of the investigations made, and a warning should be given of possible hidden trouble that the surveyor is in no position to discover. If a bressummer is involved and there are elaborate linings to window openings and decorated cornices or enrichments to the ceilings underneath, the cost of the remedial work when decay is discovered may well run into many thousands of pounds.

In carrying out a detailed structural survey it is, of course, necessary to establish whether any suspect timbers are in fact decayed. The surveyor will have to decide whether he will do less damage by making holes, say 225 x 225 mm (9 x 9 inches) in area, through the external wall, instead of taking up parquet or secret-nailed flooring. It is important to bear in mind that the sub-floor underneath may run at right angles to the external wall, or there may be two sub-floors, one at right angles to the other. In these circumstances, the area of finished flooring to be disturbed will be substantially greater than where there is only one sub-floor, with floorboards parallel to the external wall. Opening up may reveal that the bressummers or main bearings are neither timber nor steel, but cast or wrought iron, the quality of which is too uncertain to permit applying accepted engineering formulae for calculating safe working loads for the floors. Such calculations of ancient beams are of course, not the province of the ordinary building surveyor, who should advise calling in a structural engineer.

Tiled floors to balconies, and even York stone slabs, should be viewed with suspicion, because such floor coverings are not completely impervious to water. Over the years, water is liable to seep between the tiles or stone slabs, finding its way into the external wall, and setting up fungal decay in any timbers housed in the external wall. When the house being surveyed is one of a terrace it is advisable to ascertain whether the balconies of adjacent houses are still tiled or covered with stone slabs, or whether asphalt or lead has been substituted. If this is the position, it is an indication that trouble in one form or another has been experienced in some of the adjoining houses, and similar defects may exist in the property being surveyed. It is the surveyor's duty to make as thorough a search as is possible for any signs of decay in timbers in contact with, or housed in, the external wall at or below balcony level.

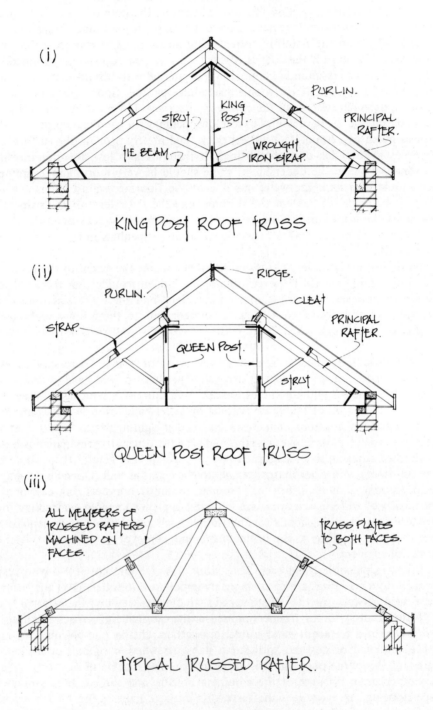

Figure 20 Roof trusses

Roof construction

The surveyor should be knowledgeable of the many different forms of construction found within roofs. The geometry of loading and stress distribution must be understood in order for the surveyor to identify potential areas of weakness.

The simplest form of roof construction consists of wall plates, rafters and a ridge board in the configuration shown in figure 17. As the span increases, collars are introduced to restrain the outward push of the roof, and the nearer the collars are to the feet of the rafters, the more effective they are in preventing this outward thrust. In the simple lean-to (mono-pitch) roof, or in double-pitch roofs for garages and outbuildings, it is not unusual to find collars on only every third pair of rafters when, provided they are of sufficient depth, they can constitute the ceiling joists. In poorly designed speculatively built, older houses, the rooms on the top floor are often partly accommodated in the roof void, with the result that the collars are pushed up nearer to the apex of the roof, progressively reducing their effective function of restraining the outward thrust of the rafters.

In buildings sixty years of age or more, it is usual to find some form of roof truss at about 2100 mm (9 ft) centres, with common rafters between the trusses. The commonest types are king-post and queen-post trusses which are used for spans up to 12–15 metres (40–50 ft), see figure 20((i) and (ii)). King-post trusses may also be found in the older, much smaller, cottage property. Such trusses called for great skill in manufacture and assembly which was mainly due to the complexity of the carpentry joints. Often the king and queen posts were further secured to the tie beams with iron straps, as were the principal rafters. In earlier examples of such trusses, the different members were tenoned into mortices in adjacent members, the tenons being secured in their mortices with timber dowels. In these older buildings, the condition of the dowels will warrant close inspection for it is likely that the dowels will have been completely eaten away by old beetle infestation. Also, the distance pieces between the inner and outer wall plates are particularly vulnerable to fungal decay, resting as they do on top of a thick wall and often under the sole of a parapet gutter.

There are usually no ridgeboards in such old roofs, rafters being halved over one another at the apex of the roof, and secured by timber dowels. Alternatively, one rafter may be tenoned into a mortice in the opposite rafter, the two rafters being secured with a timber dowel. Collars were similarly secured to rafters by dowels and, again, the condition of the dowels should receive close attention. Slate or tile battens were usually of cleft timber and slender by present-day standards, even allowing for the fact that the battens will almost certainly be oak and not softwood. Although the early craftsman were capable of doing very elaborate work extremely well, they did not always appreciate the engineering aspects of their complex designs. In consequence, the dimensions of the timbers they used were often unnecessarily generous, and not used to best advantage. In particular, the lesser dimensions of rafters and purlins were almost invariably used in depth, with the result that the roofs were not as strong as they might otherwise have been. Further, the craftsman did not appreciate the significance of knots nor the objection to using timbers containing excessive quantities of sapwood. When these roofs are inspected 300 or 400 years later, several timbers are likely to be found to have failed mechanically because of the large size of the knots or destruction of the sapwood by beetle infestation or fungal decay.

With such old buildings the roof covering of slates or tiles, unless slopes have been stripped within the last eighty years, will be secured by timber pegs, which, if of oak, are often surprisingly sound. Lime mortar 'torching' and hair plaster on the underside of the roof covering was a common method of insulating and draught-proofing, within the roof void. With the passage of time this rendering is likely to have disintegrated, adding to the dirt in old roof voids. In surveying such roof voids, often with 30 metres (100 ft) of wandering lead, the surveyor is likely to suffer considerable discomfort unless he wears a mask.

When surveying such roofs today, it will be found that time will have taken its toll. Often, substantial repairs will have been carried out by old craftsmen, without reference to structural functions of the members, with the result that positive harm may have been done. Unless a surveyor has specialist engineering knowledge, he would be well advised to confine himself to determining the extent of fungal decay and beetle infestation in the roof timbers. He should make sketches of the roof components, leaving it to a consulting structural engineer with a specialist knowledge of timber engineering to carry out a mathematical analysis of the construction, and to design appropriate remedial measures. It will be necessary to cut out lengths of seriously affected timbers, no longer capable of performing any structural function, but the expert in the engineering field will know where to introduce steel rods or straps, and even plywood gussets to restore structural strength in such a way that the essential features of the old roof can be preserved.

Packing out of old roof timbers to permit re-laying the roof covering appears to have been common practice in the nineteenth century. Unfortunately, the timber used for this purpose was often poor quality softwoods, with the result that the packing pieces are frequently found to be in a worse condition than the original timbers, perhaps 400 years old. When such roofs have to be stripped and re-covered today, repairs to the roof timbers are likely to be a large proportion of the total cost of renewing the roof covering.

The timbers that should be given the most careful scrutiny in the course of a survey of really old roofs are the wall plates, the feet of rafters, the ends of tie beams, and the distance pieces between two wall plates, particularly when these timbers are close to, or actually under, parapet gutters. Timbers of the end trusses tight to external gable walls should be individually inspected, as should the backs of wall posts of hammer beam and similar trusses. Timbers in these positions may suffer decay and often secondary death watch beetle infestation, although the remaining roof timbers of the same age are still perfectly sound. Often, when in the attic it will be impossible to make an adequate inspection of the ends of tie beams, the feet of rafters and the wall plate. However, where fungal fruiting bodies are seen on the internal face of the wall or, where probing discloses decayed timber such as at the ends of tie beams emerging from under the sole of a parapet gutter, it will be necessary to insist on bays of the gutter being lifted, to permit a proper inspection from above. Even though the original lead may be in sound condition, it is never prudent to re-lay and re-dress old lead, further adding to the cost of repair.

Another common failure of nineteenth-century restorers was to cut out lengths of decayed wall plates, building up the voids in brickwork without introducing spreaders, figure 9(iv), under the tie beams. As a result of point loads, there is a tendency for cracks to develop in the re-built brickwork. Another important aspect the surveyor should always have in mind when inspecting period properties, is the possibility that a previous owner in the course of modernising the property may have cut away some vital structural timber without taking steps to re-distribute loads adequately. It is very common to find that a purlin has been cut through to improve access to an attic without carrying out the necessary construction work to prevent the lateral spread of the roof. It is, however, perfectly possible to remove part of a vital structural member such as the lower length of a principal rafter, provided the re-distribution of loads is properly designed and configured with other timber members. Figure 21 illustrates how this was done to provide staircase access to a roof void that was to be converted into a flat. The scheme was designed by a timber engineering consultant and,

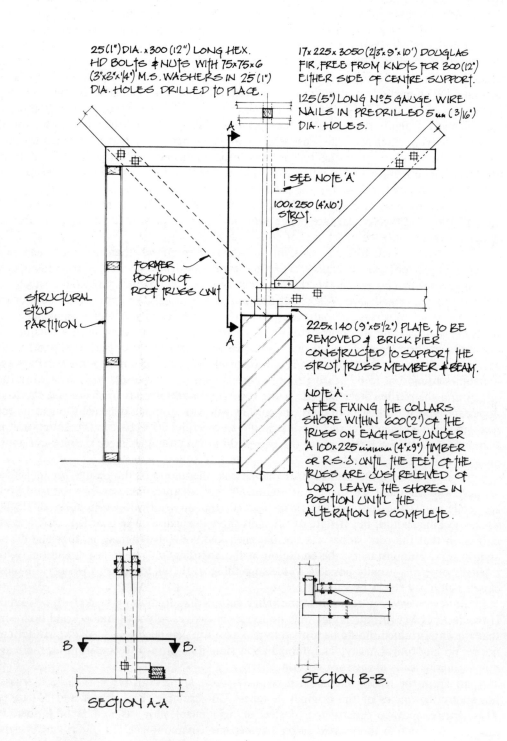

25(1") DIA. x 300 (12") LONG HEX. HD BOLTS & NUTS WITH 75x75x6 (3"x3"x1/4") M.S. WASHERS IN 25 (1") DIA. HOLES DRILLED TO PLACE.

17 x 225 x 3050 (2/3" x 9" x 10') DOUGLAS FIR, FREE FROM KNOTS FOR 300 (12") EITHER SIDE OF CENTRE SUPPORT.

125 (5") LONG No5 GAUGE WIRE NAILS IN PREDRILLED 5 mm (3/16") DIA. HOLES.

SEE NOTE 'A'

100 x 250 (4" x 10") STRUT.

FORMER POSITION OF ROOF TRUSS UNIT

STRUCTURAL STUD PARTITION.

225 x 140 (9" x 5 1/2") PLATE, TO BE REMOVED & BRICK PIER CONSTRUCTED TO SUPPORT THE STRUT, TRUSS MEMBER & BEAM.

NOTE 'A'.
AFTER FIXING THE COLLARS SHORE WITHIN 600 (2') OF THE TRUSS ON EACH SIDE, UNDER A 100 x 225 minimum (4" x 9") TIMBER OR R.S.J. UNTIL THE FEET OF THE TRUSS ARE JUST RELIEVED OF LOAD. LEAVE THE SHORES IN POSITION UNTIL THE ALTERATION IS COMPLETE.

SECTION A-A

SECTION B-B.

Figure 21 Removal of roof members

from so detailed a drawing, local carpenters had no difficulty in providing the headroom for the new staircase, which was an essential part of the conversion.

As roof span increases, longer rafters are required between ridge and wall plate. To avoid sagging in these cases, rafter depth has to be increased or the rafters have to be supported in their length by purlins. These are usually positioned midway up the roof slope but sometimes when loads are considerable or where spans are excessive, two purlins to each roof slope may be found. With purlins of adequate size and suitably spaced, simple close-coupled, double pitch roofs can be used for spans up to 9 metres (30 ft), with rafters 100, 125 or 150 mm (4, 5 or 6 inches) in depth, depending on the steepness of the pitch, and the type and weight of the roof covering. Such roofs dispense with elaborate carpentry joints being framed up with nails. In older work of poor quality, there has been a tendency for carpenters to skimp on nails thus encouraging possible collapse. In this roof type, the rafters normally over-sail the wall plates, being either spiked to the plate or birdsmouthed over the plate. Where a gutter board is fixed to these over-sailing rafters, the eaves gutter is likely to be well clear both of the wall plate and the vertical face of the wall and, in consequence, fungal decay problems at eaves level are rare with such a roof construction.

It is relatively easy for the surveyor to satisfy himself whether the roof members of such simple roofs are adequate. Together with measuring the size of roof members, it is important to check that the rafters which will be cut on the splay at the ridge are still tight to the ridge board. Also, that the purlins are straight and adequately supported where necessary by struts off partitions and that the partitions used for supporting such struts are load-bearing. If purlins are of inadequate section, or, if they are not adequately supported in their length, sagging may occur to the extent that the purlins slip off their bearings at either or both ends. As a result of shrinkage or distortion of purlins, rafters are sometimes found not to be bearing on the purlins and, in these cases, wedges between the rafters and the purlins are required. If the roof void is rectangular in shape and the roof slopes are hipped back at both ends, the purlins will be very much dependent on struts for their support and therefore, these struts should be carefully inspected. In this situation it is also important to check that where the two purlins at right-angles meet at splay-cut ends, the splayed faces are in contact.

It remains to check the condition of the battens, the nature of the fixings for the slates or tiles, whether there are soakers at abutments and whether stacks within the roof void are rendered. If the roof void is close-boarded, the surveyor will be in some difficulty when checking fixings, but if slate or tile nails show through a close-boarded roof, it is an indication that the roof slopes are not battened and counter-battened, as they should be and, in these circumstances, the condition of the boarding will need close attention. With a felted roof, it is usually possible to see something of the underside of the slates or tiles, close to chimney stacks or valley gutters.

In roofs with eaves gutters it is in the valley gutters that decay is most likely to be found. If the decay is wet rot, the attack may no longer be active. Any roof timbers and boarding that are water-stained should be probed to make certain that persistent seepage of water has not set up any fungal decay. The stained areas should be probed to confirm that there are no continuing leaks in the roof covering.

Apart from the trusses discussed previously, there are several types of framed trusses known by the names of the original designers, for example, Fink, Pratt and N-trusses. These trusses may be constructed of steel or light metal alloy, as well as of timber. In modern construction the trussed rafter has become commonplace for traditional low-rise housing. Figure 20(iii) illustrates a typical trussed rafter. These factory produced trusses are strong, light, economic in material costs and have a number of other advantages related to the ease of on-site erection. However, recent investigations have uncovered a number of defects with trussed rafter roofs which the surveyor should consider. Such defects include the lack of adequate diagonal bracing to prevent racking sideways of the trusses and the omission of galvanised steel strapping on the last two or three trusses to the gable-ends. In addition, with the risk of high humidity levels occurring in some badly ventilated roof voids, corrosion of the galvanised steel connectors can develop which will affect the structural integrity of the trusses.

Inspection of roof voids

Unless access is impossible there should be no excuse for not inspecting roof voids. Because roof voids are apt to be dirty, a surveyor may as a matter of convenience, defer inspection of them to the last, although for many surveyors the proper sequence is to commence with the roof construction, which means getting into the roof voids. Unfortunately not all are provided with access hatches, and even those that are may still defy a detailed inspection of the roof timbers if the pitch of the roof slopes is shallow. In a survey the surveyor obviously cannot have holes cut in ceilings to gain access, but in a detailed structural survey access must be obtained to all voids. Where rooms have been formed in the roof space it is important to gain access to voids behind ashlaring to inspect the feet of rafters, the wall plates and the ends of floor joists, and also to check on the condition of the plaster laths.

The information the surveyor will be seeking is the adequacy of the roof timbers, their condition and whether the roof has been properly framed together. He should have sufficient knowledge to be able to determine whether the roof timbers are of softwood or hardwood and, if the latter, whether the timber is oak, elm, or sweet chestnut, or a species he is unable to identify.

As more and more insulation is placed in roofs there is an increased likelihood of condensation occurring. Where insulation is placed between the ceiling joists, roof voids will be cooler and, in cold weather, water vapour from below may form droplets on cold surfaces within the void. To overcome the problem it is essential to ventilate the roof, usually via grilles set in the soffit board. The surveyor must ensure that ventilation is adequate in these circumstances and check that grilles are not blocked by the over-zealous application of insulating materials. *BRE Digest 270* 'Condensation in insulated domestic roofs' discusses the problem of condensation and offers suggestions on reducing the risk of it occurring.

While in the roof void he should note the condition of any electrical wiring lying on top of the ceiling joists and, particularly, whether the junction boxes have protective covers, although, of course, he will leave it to an electrical engineer to test the wiring. If lagging to pipes and cold storage tanks is of felt, a warning should be given that this fosters the breeding of clothes moths and carpet beetles and should be replaced by a more effective inorganic lagging. Storage tanks should be inspected internally for evidence of rust, as should the tappings. In old roof voids it is often advisable to recommend the clearing away of accumulated debris and of flammable materials such as straw and twigs brought in by birds. If the ceilings of the rooms underneath are of lath and plaster, then the condition of the plaster key should be checked.

When in the roof space of terrace or semi-detached houses, the surveyor should check the condition of the party wall. This wall should be taken up through the roof void and, often in older properties, may pass right through the roof between each dwelling to rise well above the roof covering. Such party walls are not always present, even in good-class semi-detached houses, particularly those built just prior to the close of the nineteenth century. The party wall is essential in reducing the risk of spread of fire from one dwelling to the next. For this purpose the wall should be non-combustible and all junctions between the wall and roof should be fire-stopped.

It is important to discover whether there is any beetle infestation in the roof timbers, what beetles have been at work and whether the infestation is still active. Even if the attack is quite extensive, provided no real structural damage has occurred, there will be no need for any remedial work where infestation is no longer active. The search for evidence of attack, particularly of the common furniture beetle, is likely to take longest—size for size of roof void—if the house is only ten or fifteen years of age. This is because there will not have been time for frequent re-infestation to have occurred, and there may be very few flight holes, which may take much searching to find. The places to search initially are the framing to the access hatch and any boarded areas of the roof void where discarded articles have been stored, particularly plywood chests and wicker-work furniture, linen baskets and the like. A warning should be given to the client that storage of such articles is inadvisable. In his survey report, it is important for the surveyor to state precisely how thorough his search has been. A close inspection of all faces of every roof timber could well demand the greater part of a day in the roof void of a three-bedroom terrace house, less than thirty years old. This is the only method open to a surveyor required to give an absolute guarantee that there is not a single flight hole in the whole of a roof void. With experience a surveyor will be able to assess the position with something less than this. During his inspection he should shine his torch up every rafter, and along each purlin and ceiling joist. In the roof of a house twenty or more years old, he will probably find some evidence of furniture beetle attack within the first few minutes of his inspection of the roof void. He can then much more usefully employ himself in his client's interest in assessing whether attack is of any practical significance and what, if any, remedial measures should be recommended.

If there is evidence of widespread attack in an old roof, the surveyor should look, when crawling across the ceiling joists, for beetles lying on top of the plaster ceiling. If only remains of beetles are found, or beetles that have obviously been dead for some years, it is most unlikely that there is any continuing active infestation. The appearance of frass in a roof void may be misleading, for example, it may still be in little piles and spilling out of flight holes and comparatively light in colour, some years after the last adult beetle emerged. When there is real doubt as to whether there is an appreciable amount of continuing active infestation, it is advisable to recommend that the roof timbers and the top of the plaster ceilings should be cleaned down with a powerful vacuum cleaner, which will make it easy to determine the extent of continuing active infestation in subsequent flight seasons.

House longhorn beetle infestation may be important, but experts know of no case of continuing active attack in the original roof timbers of any building more than fifty years of age. In the Home Counties, where attack is known to be endemic in houses from the 1930s onwards, a very thorough search for house longhorn infestation is obviously necessary. Rafters on each side of chimney stacks, with flues in use, will often provide the first evidence. The warmth from such flues creates appropriate habitat conditions for hatching and development of house longhorn beetle larvae. It is thought that adverse weather conditions in certain years, coupled with the long life cycle of seven to ten years, explain why attack has often died out in older buildings. Since the larvae feed on sapwood they are inevitably near the surface of the timbers in which they are feeding, and hence are exposed in hard winters to temperatures too low for the larvae to complete even one life cycle. Moreover, because of the long life cycle, male and female beetles may not emerge in the same year, making re-infestation impossible.

Holes of much the same size as the flight holes of the common furniture beetle may in fact be the galleries of the pin-hole borer which are of no consequence, since infestation occurs in the forest, not after conversion of logs to sawn timber.

In assessing the practical significance of any beetle infestation, but particularly attack of the common furniture beetle and the house longhorn beetle, what matters is the amount of sapwood in the timber. House longhorn beetle infestation is confined to the sapwood, and although common furniture beetle attack is not, the amount of infestation in sound heartwood is often negligible. In houses built before the First World War, the amount of sapwood in carcassing timbers then available was so small that widespread furniture beetle attack is unlikely to be of any structural significance. This is also the case even if the plaster laths at the back of the ashlar walling are riddled with flight holes. In the lesser period houses and cottages where round hardwood poles were used for rafters and ceiling joists, it is probable that the sapwood will have been completely eaten away, first by *Lyctus* infestation, and later by furniture beetle attack. In these houses, if attack is severe, there will be insufficient timber left to perform its structural function and hence the proliferation of sagging roofs and plaster ceilings so often seen in such properties. In these circumstances an insecticidal treatment is pointless and renewal of ceiling joists and rafters would be advisable. In most properties, instead of spending a large sum of money on attempting to eradicate a possible small amount of continuing active furniture beetle attack, the client should be advised to treat any valuable pieces of furniture prophylactically.

Active *Lyctus* infestation should not be a problem in roof timbers today because hardwoods generally, but oak in particular, are rarely if ever used in roof construction. If repair or renewing in oak or similar *Lyctus*-susceptible timbers is specified, for example, in the roofs of buildings of historical interest, the timber used in repairs must be absolutely free from sapwood. In old buildings where the roof timbers are of oak, evidence of death watch beetle attack may be found, and a thorough search should then be made for fungal decay, which is likely to be of greater structural importance than the secondary death watch beetle infestation. Death watch beetles are much more sluggish than furniture beetles and apparently dead beetles may begin to move after being placed in the palm of one's hand for a minute or so. What is of assistance in assessing the extent of continuing attack is the number of live beetles found, assuming the survey is being carried out during the flight season, which is usually late April or early May. In conditions that are marginally suitable for the death watch beetle, the life cycle may be very prolonged so that the occasional adult may emerge long after an infestation has ceased to be of any practical significance.

Because it was quite common in medieval England to demolish buildings and to reuse the timber, death watch beetle flight holes are often seen without there being any traces of frass. If there is continuing active attack a sharp lookout should be kept for the steely blue beetle, *Corycorinetes coeruleus Deg*, a predator on the death watch beetle. It is exceptional not to find this predator if there is any significant continuing active death watch beetle infestation. When surveying a large high room such as a college dining hall and where death watch beetle infestation is known to exist, the extent and location of continuing attack can often be usually assessed by crawling over the floor during the flight season. It will usually be possible to find some beetles, apparently stunned, on the floor, and of course if they are at all numerous it will be necessary to obtain a tower scaffold to permit examination of the roof timbers at close quarters.

The only other wood borer that the surveyor may encounter in attics is the wood-boring weevil. This however is unlikely for the habitat of this beetle is typically basements and poorly ventilated ground-floor floors, not roof voids. Wood-boring beetles are discussed in more detail in appendix A.

Fungal infection of roof timbers

Fungal attack is likely to be a more serious matter than beetle infestation because the strength properties of attacked timber are seriously reduced. Moreover, *in-situ* chemical treatments will not suffice to deal with fungal infection. By the time fungal attack is discovered, actual structural damage is likely to have occurred, necessitating renewal of some timbers. It is the ancillary work that is usually the most costly part of the remedial measures, particularly when elaborate plaster cornices and panelling are involved.

Fungi are plants that differ from other members of the plant kingdom in that they do not manufacture their own food requirements, but feed on the tissues or the cell contents of other plants and living organisms, including man. Fungi responsible for decay in timber, feed on the cell wall substances of which wood consists. Five essential conditions are necessary for fungal attack to occur:

(a) A source of infection, that is, spores (the equivalent of the seeds of higher plants) or *mycelium* (the vegetative parts of fungi).
(b) Food material, which means most wallpapers or timbers, although some species are more resistant to decay than others, and the sapwood of all species is more readily attacked than the heartwood.
(c) Oxygen, which is available from the atmosphere.
(d) Suitable temperatures, as fungal growth is either at a standstill in unheated buildings during winter or it is extremely slow, becoming progressively more active as temperatures rise in the warmer months of the year.
(e) Moisture, which is as necessary for the so-called dry rots as it is for the wet rots.

On finding *mycelium* or fruiting bodies, the surveyor needs to direct his attention to locating the source of moisture that has made attack possible. In roof voids defects in flashings, valleys and roof coverings are the most likely sources of water that give rise to fungal attack. At wall-plate level and in the vicinity of parapet gutters, breakdown of the rainwater disposal arrangement is the most frequent cause of trouble. At lower levels downpipes, string courses and defects in the fabric of the building provide opportunities for ingress of water. Pipe runs, plumbing generally and persistent condensation, provide water or water vapour within the structure. At ground-floor or basement-floor levels, the absence of effective damp-proof courses and adequate oversite concrete result in the build-up of high relative humidities which, in turn, raise the moisture content of wood above the critical minimum necessary for fungal attack.

It is important to be able to identify the fungus, that is, whether dry rot (*Serpula* or *Poria*) or a wet rot, which is nearly always *Coniophora* in softwoods and *Phellinus, Poria spp.*, or *Coniophora* in oak. *Serpula* is rarely found attacking oak. In very wet conditions, as under leaking parapet gutters, *Paxillus panuoides Fr.* may occasionally be found. The fruiting bodies of this fungus are rather attractive in appearance, being bell-shaped, olive-green in colour with deep gills on the under surface. It is sometimes a little difficult to distinguish *Serpula* from *Poria* in the early development of the fruiting body. If the surveyor takes a piece of the fruiting body or some thick strands of *mycelium* back to his office, and exposes them to the air overnight, identification should not be difficult by the time he comes to write his report the next day: *Serpula* will have become brittle, whereas *Poria* will have remained leathery. The common wood-destroying fungi are described in greater detail in appendix B.

In practice the surveyor will primarily be concerned with *Serpula, Poria* and *Coniophora* in softwoods, and *Coniophora* and *Phellinus* in hardwoods, and it is essential for him to familiarise himself with the distinguishing features for identifying these four fungi. He should, however, be able to recognise elf cups (*Peziza* sp.) and a species of inky cap (*Coprinus* sp.) to be found growing on plaster ceilings that have been saturated, following defects in plumbing or frost damage. Identification is important because, although these two fungi do not attack timber, they are an indication that dangerously damp conditions exist, and rapid drying out is called for if serious outbreaks of dry and wet rot are to be avoided.

Serpula and *Poria* can be identified from their *mycelium* or fruiting bodies, *Coniophora* from its *mycelium* (a surveyor is most unlikely to find the fruiting bodies of *Coniophora*) and finally, *Phellinus* by its fruiting body. Attack may well be present without there being any visible *mycelium* or fruiting bodies, or even a characteristic musty smell, and this is where the surveyor's expertise is so vital to his client. It is impracticable to probe every square millimetre of timber in a building and in fact, this is not necessary, but it is essential to investigate wherever attack is to be expected.

The presence of advanced decay should present no problem but it is the early stages of attack, or the existence of deep-seated attack, that give rise to real difficulties. Often the surveyor will have no more than symptoms to go on, for example, damp stains on walls or ceilings, curved surfaces of skirtings, door linings and floorboards and, of course, the various signs that he will have noted in the course of the examination of the building externally, that provide evidence of water penetration of the fabric.

Flat roofs

Flat roofs have been used successfully for over three centuries for large and small houses. Unfortunately, flat roofs erected over the last fifty years have attracted a very poor image because of a large number of factors such as the use of innovative but inadequate designs and the incorporation of untried and untested materials. Little was known about the problems that could be met with such inadequate designs but as time has moved on, designers have acquired more and more knowledge. However, although such technical knowledge is readily available today, developers and designers of properties, in all price ranges, still continue to make the most elementary mistakes in design and construction with the result that serious defects in flat roofs are often occurring within eighteen months to five years of completion of the building. In fact, there is more justification for suspecting trouble in flat roofs of developments constructed after the Second World War than in those of period houses and public buildings 200 to 300 years old or more. The reasons are many but include the following:

(a) Modern planning of the accommodation under flat roofs.
(b) Labour-saving appliances such as washing machines, tumble driers, etc., which produce vast quantities of water.

(c) The habits of occupants, often dictated by economic considerations.
(d) The absence of open fires with their invaluable flues to provide some ventilation.
(e) Inadequate design such as the specification of inferior materials both for the construction and the roof covering and also, the lack of any fall to the flat roof to aid the removal of storm water and the omission of vapour checks, ventilation and insulation.
(f) Poor workmanship.

The original coverings of flat roofs of buildings a hundred years of age or more will be either lead or copper. The thick cast lead of sixteenth and seventeenth-century buildings is often still sound today, for example, that used on the large expanse of flat roofs of Wollaton Hall, Nottingham.

The 34 kg/sq. metre (7 lb) and 39 kg/sq. metre (8 lb) milled lead of the Victorian era can be expected to have a useful life of eighty to a hundred years, assuming it was properly laid in the first instance and that it has been well maintained. Copper is used in thinner sheets and is more prone to suffer damage from impurities in the atmosphere and, therefore, it usually has a shorter life than lead. As for the roof timbers in flat roofs, until well into the eighteenth century those of the large private houses and public buildings will be of hardwood and not softwood, and the walls in which the ends of the timber are housed are likely to be appreciably thicker than in more modern buildings, hence the timbers will have been less exposed to the hazards of decay.

Today, except in important buildings, lead and copper are little used because of their relatively high cost. Towards the end of the nineteenth century a cheap alternative, zinc, was common in less expensive suburban properties. Unfortunately, zinc has also been used abundantly in repairs, particularly for small areas of flat roofs, parapet gutters and dormer cheeks. The life of zinc may be as little as twenty years and, as has already been indicated, its use in repairs should put the surveyor on his guard as there may well be other and more serious defects in buildings as a result of such economising.

The most recent metal to be used for roofs is aluminium which has the great advantage of light weight and relatively low cost, but can give rise to condensation problems if suitable precautions have not been taken. On clear cold nights the drop in temperature of aluminium roofs, as a result of radiation, can be as high as 5°C, causing formation of ice both on the top surface of the metal, and on the underside. When temperatures rise during the following day, dripping of water on to ceilings can be severe, causing dampness in the building. Ventilation of the roof space cannot alone prevent condensation on the underside of the roof sheeting when the roof is radiating heat to a cold night sky, but it is useful in getting rid of water by evaporation during the day. A combination of vapour barriers, ventilation and insulation may suffice but careful detailing is necessary to prevent interstitial condensation forming.

Since the turn of the century asphalt has been widely used and, if it is good-quality mastic asphalt of adequate thickness, and properly laid, it is a satisfactory roof covering and one that can be easily repaired when damage occurs. In the early days, asphalt was laid direct on to the roof and much blistering and cracking resulted. Today, asphalt is laid on an isolating membrane in compliance with *CP 144*: 'Roof coverings Part 4: 1970 Mastic asphalt', and blistering and cracking from this earlier bad practice has been eliminated.

In the last twenty years probably the most frequently used material for flat roofs is bituminous felt. Provided it is three-ply felt, conforming to *CP 144* : 'Roof coverings Part 3: 1970 Built-up bitumen felt', and is laid strictly in accordance with the recommendations of this Code of Practice, it is a satisfactory roof covering with a thirty year useful life, if properly maintained.

Another development in flat roof construction was the use of the thermal insulation boards to replace the traditional timber decking. In earlier work the two most frequently used materials were Stramit board and woodwool slabs. Stramit board is compressed straw which was supplied in standard-size sheets wrapped in a paper envelope, the top side of which was impregnated with bitumen to protect the board during laying. The top surface was no more than a water-check, and the roof was to be properly sheeted should rain occur before the first layer of felt could be applied. Unfortunately, this did not always happen which then resulted in early failure of the decking material. With Stramit, it was also important for the board to be laid in the direction recommended by the manufacturer, namely, with the straws at right-angles to the supporting joists. Further, when it was necessary to cross-cut a board, the raw ends were to be sealed with tape supplied by the manufacturer. Being an organic product, it was vital for the roof void to be ventilated to the atmosphere, and for a vapour barrier to be applied to the warm side of the ceiling. Failure to take these precautions often had devastating consequences. Condensation within the roof void and on the underside of the Stramit created ideal conditions for fungal and mould growth within the straw, which quickly reduced it to humus, destroying its structural properties as a suitable base for the usual three-ply roof covering.

Woodwool is an open textured board comprised of wood fibres bound together with Portland cement. Woodwool, although predominantly organic, is not normally susceptible to fungal attack, but even so, ventilation of the roof void is necessary to protect the roof timbers. This material has been, and still is being, used successfully as a decking material for flat roofs.

Another alternative to timber decking is chipboard, but this material does have its problems in that it shrinks and swells appreciably, particularly in its thickness, with changes in the temperature and relative humidity of the atmosphere. Moreover, cyclic changes in moisture content are likely to cause a reduction in its inherent strength properties. Care should be exercised when inspecting such materials, and any bowing or saturated areas should be highlighted for further investigation.

The most frequent cause of early deterioration of asphalt and felt-covered roofs, where the construction is not at fault, is failure to protect the decking against inclement weather while laying is in progress. With timber decking and three-ply felt, it is important that the first layer of felt is nailed and not stuck down, and the decking on which the felt is laid must be in a stable condition. Local stretching of felt, by as little as 2 per cent, can render it porous without there being any visible splits in the felt.

The suitability of the material used for covering flat roofs, including correct laying and its present condition, are points that the surveyor has to assess from his visual inspection. With lead, copper and aluminium, it is important that bays should not be more than 825 mm (2 ft 9 inches) between rolls and not more than 2400 mm (8 ft) between drips. Where larger bays or longer lengths in gutters have been used, there is likely to be evidence of much repairing of old splits. It is exceptional to find properly burned-on patches in recent · repairs, the usual method being a soldered patch, which is very liable to split again after a few years. It is usually possible to ascertain the thickness of the metal at a drip. In the absence of positive signs of deterioration, the age of a building provides a useful yardstick for assessing the further life of the roof covering, bearing in mind that even 34 kg/sq. metre (7 lb) milled lead eighty years old will be approaching the time for renewal.

If the timber decking shows through the metal covering, or the roof is springy to walk on, replacement of the roof covering and boarding underneath will be essential. In addition, where lead roofs or gutters are found to be coated with bitumen or felt it is an indication that early renewal of the lead will also be necessary. Furthermore, a coating of heavy-consistency bitumen found in such places should be regarded as giving a further serviceable life of not more than five years. When properly treated by a specialist firm with hessian and bitumen, some firms will give a guarantee of ten years, provided subsequent dressings of bitumen are applied every three years.

Lead treated by any of the methods described above should give rise to doubt not only about the condition of the decking, but also of the roof timbers themselves, and this point should be made in the surveyor's report. Further, in drawing attention to splits in lead or copper roofs requiring patching, he should stress the importance of employing only really competent firms for effecting repairs because of the potential fire hazard of contractors' burning equipment. Fire extinguishers should always be to hand when repairs to lead flats and gutters are being undertaken.

With copper roofs, it is important to check whether rainwater is being taken away in cast iron pipes for, unless these pipes have been coated internally with a bituminous paint, they will have a comparatively short life. The tear marks on lead upstands caused by dripping water from the eaves courses should be checked to ascertain how much metal has been worn away. If this is appreciable, the metal may be paper-thin and probably full of pin holes.

Since asphalt and felt are not laid in bays or with drips, the surveyor has only the superficial appearance of these roof coverings to assist him in assessing their future life. The appearance of creases in asphalt and blisters in felt may be a warning that early replacement will be necessary. Bearers for water tanks can also be a source of trouble when they rest directly upon the roof covering, but most damage arises from the use of such flat roofs as gardens, when the legs of chairs and stiletto heels frequently puncture the covering.

Disposal of rainwater from flat roofs is very important, particularly where falls are negligible. In older buildings there are usually parapet gutters with outlets through the parapet wall discharging into hopper heads. With smaller areas there may be a slope in only one direction, discharging into an eaves gutter. There is an increasing tendency today to return to outlets discharging into downpipes, with the attendant risk of the outlet becoming blocked, particularly where there is no hopper head. Such outlets should be protected by wire cage balloons, which require frequent inspection in autumn to ensure that the outlets do not become blocked with leaves. Advice on such points should always be given in the survey report. The surveyor should be particularly suspicious of flat roofs to bay windows, particularly when the bay is not carried up to the full height of the building, for example, only to first-floor level. In such circumstances the external wall of the upper storeys will be built off a bressummer, which in many older buildings is likely to be of timber. Where the flat roof over the bay slopes towards the external wall, there is a risk that rainwater will penetrate under the cover flashing to run down behind the upstand of the roof covering, and on to the bressummer below. Repeated seepage of rainwater in this manner creates ideal conditions for the development of dry rot in the bressummer. There are often no visible signs of any trouble from within the house unless the decay has reached an advanced stage. The surveyor should be aware of such a risk and must sound a warning in his survey report as replacement of a bressummer will be a costly operation.

As has already been stated, the surveyor is rarely able to check the construction of flat roofs, other than those of outbuildings and garages that have no ceilings underneath. All he can normally do is to draw inferences, based on his professional skill, from the condition of ceilings and walls under flat roofs and of other parts of the building, and give suitable warnings in his written report. He may be able to estimate the depth of joists at access hatches, but he should have been able to form as useful an assessment of the adequacy of the timbers in walking over the roof. Obviously the condition of ceilings immediately under flat roofs is important, but if these have recently been re-decorated, defects will have been covered up. Recently re-decorated ceilings should always arouse the surveyor's suspicions. Other clues may be provided if the ceilings under hollow flat roofs are of different construction from the ceilings in other parts of the building, for example, plasterboard or similar materials under flat roofs where ceilings elsewhere are of lath and plaster. It is always worth looking for electric light fittings recessed in ceilings under flat roofs, removal of which may give some indication as to the roof construction and condition of the timbers.

Faults in construction and design are so prevalent in flat roofs of modern blocks of flats and maisonettes that every effort should be made to see the original plans and specification. The commonest faults are:

(a) Want of ventilation of the roof void.
(b) Use of untreated timber or, at best, timber that has only received either a dip or brush treatment, instead of being pressure-treated.
(c) Omission of a vapour barrier on the warm side of the roof.
(d) Lack of sufficient fall to the roof to remove storm water.

Given these shortcomings in construction, condensation troubles can produce devastating consequences in a very short space of time. Lack of ventilation of the roof void, together with a vapour barrier on the cold side of the roof and roof timbers not pressure-treated, should be regarded as sufficiently serious shortcomings to necessitate advising against purchase. When there have been previous occupants, a careful search will usually reveal evidence of condensation having occurred in the past. Condensation problems stem from the want of ventilation in the roof void and modern planning, such as the open plan where there is no permanent division between the kitchen and the living room.

The appliances in an occupied flat or maisonette should be noted, as gas cookers, washing machines and tumble driers are a frequent cause of condensation problems, particularly where fanlights or permanent ventilation to the outside have not been provided in kitchens and bathrooms. The method of heating is also an important factor. All too often there is only partial central heating perhaps confined to the living room and corridor and sometimes in the kitchen with only electric fires on the full domestic tariff in the bedrooms, and probably no form of heating in the bathroom or WC. If the heating system is of the ducted air type, the location of the inlet grille should be carefully considered. It is not unknown in this form of heating for the inlet grille to be in the kitchen. This will result in warm wet air being delivered to the rest of the accommodation, causing condensation on cold surfaces such as window reveals. If such accommodation has been occupied by couples out all day, the likelihood of a past history of persistent condensation is all the greater. Each adult releases one pint of water in breathing every ten hours. Further, in the morning, more water vapour is released in cooking and bathing, which does not condense while the heating is in operation. If however the heating is turned off on leaving the flat, and for security reasons windows are shut, condensation occurs as the temperature falls.

In the evening before there has been time for the accommodation to become adequately heated, more water vapour is produced when cooking the evening meal and in any clothes washing that may be done then. Thus further condensation is likely to occur. Where oil burners are used, there can be devastating consequences as heavy mould growths may develop and pools of water may accumulate against external walls in unheated rooms. This is because the burning of paraffin results in the release of water vapour which is equivalent to the volume of paraffin burned.

If the flat roof is of timber construction, unventilated, with no vapour barrier, and the timber has not been pressure-treated, condensation can result in serious decay in the roof timbers in as little as eighteen months. In one such case, mould growths in the flat roof over an operating theatre were so severe within a year of construction as to necessitate running the pressurising plant night and day, instead of only when the theatre was in use, until the roof could be stripped and the defects in design made good. It is for this and similar reasons that the surveyor when carrying out a reconnaissance survey of flat roofs on a large number of dwellings such as blocks of flats, should pay particular attention to any evidence of a past history of condensation or rain penetration, which may already have set up decay in the roof timbers.

Condensation problems from the same causes are likely to arise in flats with solid floors and roofs, and although structural damage may not result, it can render the accommodation virtually uninhabitable. Condensation in dwellings is discussed in greater detail in *BRE Digest 110* 'Condensation', *BRE Digest 180* 'Condensation in roofs', and *BRE Digest 218* 'Cavity barriers and ventilation in flat and low-pitched roofs'.

7 Joinery and Finishes

It is common when inspecting houses more than 10–15 years old to find a number of defects with the timber joinery. The quality of such joinery, particularly that produced shortly after the Second World War and up until the mid 1970s, was very poor. Failures were often blamed on one or a combination of the following factors:

(a) Inadequate protection given to joinery during the construction stage.
(b) The use of unsuitable softwoods.
(c) The use of timbers with large amounts of sapwood.
(d) The absence of effective preservatives.
(e) Inappropriate design details.
(f) Poor workmanship including the application of preservatives.
(g) The over-reliance upon mastic sealants around timber frames.

Defects during construction

Protection of joinery during the construction stage is very important, particularly if this period is prolonged. If windows and doors frames are built in as external walls are raised, external painting is often deferred for a year or more on large building projects, and the priming coat applied to the timber units before they leave the joinery works does not provide sufficient protection for such long-term exposure. Some Local Authorities now use inverted polythene envelopes to enclose windows built in as the work proceeds, but it is preferable for such joinery to be glazed and undercoated prior to despatch to the site. If external joinery is fixed in polythene 'envelopes', it is important to ensure that these are completely removed, otherwise water may find its way in later behind the remains of an envelope. Glazing and undercoating prior to despatch are particularly important if glazing incorporates glazing beads as these are not infrequently pinned in position before priming, with the result that the surfaces covered by them are not protected by the primer. After say a year's exposure to the weather, bottom rails are likely to become saturated, and subsequent glazing and painting seal in the moisture, creating ideal conditions for the development of wet rot unless the frames have been pressure-impregnated prior to manufacture.

Survey of external joinery

With decay in external joinery, it is all-important to determine whether the problem is one of ingress of water or of condensation. A logical method of inspection must be followed if the surveyor wishes to identify the real cause of timber decay. The surveyor would be wise not to jump too readily to an early conclusion. If the extent of the property is large or if there is, for example, a large number of dwellings such as 100 or more flats in several separate blocks, then the surveyor may well modify his conclusions more than once in the course of the survey. Such surveys may well take several days to complete. Data for each flat should be collected to an identical logical pattern, for example:

(a)   length of tenancy;

(b)   number of occupants and age of children;

(c)   aspect of the different rooms and location in relation to cold areas such as open corridors and stairs;

(d)   equipment in the kitchen such as gas or electric cooker, appliances (washing machine, spin drier, tumble drier, clothes boiler, airing cabinet);

(e)   form of heating, unit heaters and whether gas, electricity, or oil heaters, and if partial or complete central heating is provided, the extent to which this is used;

(f)   habits of the occupants, family out all day, or wife and children in occupation during the day;

(g)   frequency of bathing, and whether night or morning (the not infrequent habit of bathing with the door open naturally aggravates condensation);

(h)   nature of furnishings, for example, fitted carpets or rugs and impervious floor coverings;

(i)   location of members affected with decay;

(j)   type of joinery material;

(k)   type of construction including whether cold or warm roof;

(l)   form of insulation (if any), cold bridging, etc.

If these data are typed in tabular form, a distinctive pattern may well emerge, indicating that the problem is related to the design, the equipment in the flats, the habits of the occupants or the quality of the material or construction.

An example of the magnitude of the problem was revealed in one such survey, when serious decay in purpose-made joinery was found. This occurred in several blocks of flats on an exposed site on the coast, within six years of completion. The timber was untreated European redwood. It was found that some members of 94 per cent of the windows on the exposed south and west elevations were decayed, but only 15 per cent of those on the north elevations. In this case the problem was mainly caused by ingress of water, although condensation was no doubt a contributory factor to the decay on the north elevation. It was found that air of high relative humidity produced in the kitchens and living rooms (where temperatures are likely to be higher than in bedrooms, bathrooms and WCs), drifted to the colder parts of the flats, which were the rooms on the north elevation, where the windows acted as dehumidifying units.

Where the external joinery is of considerable age, some wet rot, particularly in tenons, is likely to be encountered. The surveyor should be on the look-out for evidence of piecing-in of members and the reinforcing of angles with iron brackets. Such repaired joinery may still have an appreciably further serviceable life, but it is important to record the extent of repairs that have been carried out in the past.

Windows

The type and condition of the external sills should be noted, and the material of which they are made, for example, timber, stone, reconstituted stone, bricks or tiles covered with rendering, or creasing tiles. In all but the last, which have no throats, it is important to check that there is an adequate throat which has not become filled with paint over the years. If sills are cracked, the condition of window boards and the wall plaster and skirting under the window should be carefully examined. If the face of the wall under the window is covered with a material different from the wall face elsewhere in the room, for example, with hardboard or plywood, it is often a clear indication that water has been getting in through the cracks in the sill, and the plaster wall face will almost certainly be perished. In older houses, where grounds for window boards are frequently 75 x 50 mm (3 x 2 inches) or more in section, fungal decay in the grounds is often caused by defective external sills, and this decay may spread to panelling under windows.

The surveyor must be able to differentiate between softwoods and hardwoods, and to recognise oak from other hardwoods. Ability to identify the timbers used for external joinery is of particular importance today. Much of the trouble with older joinery stems from the amount of sapwood that was accepted in the joinery grades of European redwood and similar timbers. To reduce the incidence of knots to the limits laid down in past British Standards, it was necessary to accept timber containing a high proportion of sapwood, and the sapwood of European redwood and similar species was and still is perishable. Moreover, unless the timber was kiln-dried prior to manufacture into joinery, it was likely to have a moisture content of 18 to 20 per cent, or even more. This often resulted in surface cracking and opening of joints after assembly. Avoidance of trouble called for the use of timber adequately treated with appropriate fungicides, and preferably by a pressure process, after which the timber should have been kiln-dried to 15 per cent moisture content. If the timber is European redwood the joinery should be thoroughly inspected by probing, particularly at joints. For this purpose an electrician's small screwdriver will be found most useful.

Even joinery complying with current Codes of Practice and Specifications has not always escaped early fungal attack and, where architects have produced their own details, without fully appreciating the reasons for certain important aspects of technical design, the results can be devastating. A built-up transom, assembled only with screws, and without taking the precaution of priming all the members separately with a good-quality primer prior to assembly, is especially vulnerable to subsequent fungal attack. Such designs as large picture windows encourage condensation, and V-jointed horizontal cladding is particularly prone to early decay. The risk of decay appears to be particularly great in modern blocks of flats where architects have adopted an open plan, or a design with kitchens communicating direct with living rooms. In such circumstances, in addition to the hazard from exposure to the weather, there is a considerable risk of persistent condensation giving rise to fungal decay on the inside of window members, particularly if the back putty is defective at the base of the glass, serving as a trap for condensed moisture. Sills, the bottom members of frames, the horizontal members of sash and casement windows and built-up transoms are most likely to be affected. Decay is almost invariably a form of wet rot, and to this extent there is less risk that infection in the joinery will have spread to roof and floor timbers.

Openable windows are often the only method of ventilating habitable rooms and under current Building Regulations minimum sizes of openable area are stated in relation to the floor area. The surveyor should bring to the attention of his client the need for ventilation where this is inadequate, particularly if fuel is burnt in the room such as in open fires or boilers.

The amount of natural lighting is also important and current regulations require a minimum window area, again related to the floor area.

Doors

The type and condition of doors, thresholds, sills, linings and frames should be noted in the report. Rotten members should be probed to reveal whether fungal or beetle attack is taking place. If fungal attack is present it is almost invariably of the 'wet rot' type. When inspecting decaying doors the usual reasons for early failure are inadequate detailing, lack of projecting sills, economies of construction such as minimum timber sizes, ineffective preservatives and poor seals between glazing panels and stiles and rails. Also, it is not uncommon for builders to use internal quality doors as external doors, relying on a few coats of paint to protect the surface.

Ceilings, cornices and picture rails

Ceilings of houses built since the late 1930s are almost always of plasterboard, and because the builders have usually been content only to scrim joints and apply one skim coat, cracks along the joints between two sheets of plasterboard are a very common defect. In earlier buildings, the ceilings should be of lath and plaster, but if they are of other materials, for example, plasterboard or some form of insulation board, the construction should be commented upon in the survey report and an indication given as to the probable explanation. If the survey is carried out after a series of wet years and the property has been very thoroughly redecorated, renewal of ceilings in different materials may be the only clue to a past history of settlement damage. The explanation may be old war damage, but it can be an indication of previous settlement problems, particularly settlement damage caused by tree root action in a shrinkable clay soil.

The key of lath and plaster ceilings should be tested along the line of any cracks, for if the plaster moves up and down on either side of the crack, it is an indication that much more than stopping of cracks will be called for when the ceiling is re-decorated. Sagging ceilings on the top floor of period properties should always be commented upon, particularly if, as is often the case, the surveyor has not been able to get into the roof void. It will often be found that large areas of ceiling can be moved up and down, and it is imperative to recommend that such defective ceilings should be taken down because falling plaster upwards of 25 mm (1 inch) thick could cause serious injury. The client should be warned that it may be necessary to renew the ceiling joists too, as in such old properties these are likely to be only half-round poles containing large amounts of sapwood, which will have been eaten away by old beetle infestation. Patches in ceilings on the top floor are often evidence of ingress of water, either rainwater through the roof covering, defective flashings, or leaks from pipe runs or storage tanks. These observations should be collated with any notes on the roof voids and roof coverings.

Cornices and decorated ceilings should be closely inspected, and it is advisable to check that other rooms of equal status on the same floor have similar features. The finding of simple cove cornices in one room where most other rooms in the house have more or less elaborate, moulded cornices, should put the surveyor on enquiry. There could be many explanations, for example, war damage, settlement damage of one kind or another, or evidence that there has been dry rot in the roof timbers above. The last explanation is a likely one if there is a parapet gutter above the ceiling. The cost of reinstating an elaborate moulded cornice along one wall is usually appreciably greater than the cost of taking down the cornice all round the room and substituting a plain cove cornice. If the surveyor has any doubts he should return to the roof void to check whether there are any indications of old fungal attack. Elaborate cornices or margins to ceilings may be in plaster and not timber, but if of timber and under the roof, there is always a risk that the timber members may contain fungal decay or suffer wood-borer attack.

Picture rails can also provide evidence of the existence of furniture beetle infestation. As already stated, a painted finish provides some protection against the egg-laying of wood borers, so the top edge of picture rails should be inspected because it is often the only unpainted timber in the room.

Skirtings

Skirtings should be closely inspected, for like cornices and decorated ceilings, the finding of simple skirtings in one room where most other rooms in the house have more or less elaborate skirtings can indicate some past defect. Explanations for such variations could be war damage, settlement damage of one kind or another or, more often than not, evidence that there has been dry rot or severe beetle infestation at some time. Clues like this should demand closer inspection of the skirting and floor.

Glazing

All glazing should conform with the recommendations set out in relevant British Standard Codes of Practice, but all too frequently in large-scale projects where cost has been a dominating factor, or in a factory assembly of whole wall units, puttying can leave much to be desired. A common failure with external glazing beads is the dry bedding of beads and, in factory-assembled units in particular, the omission of setting blocks and spacers. If external glazing beads are merely fixed with pins, a gap between the glass and/or putty is likely to develop, and rainwater will find its way under the glazing beads, saturating the base of the rebate, thereby creating ideal conditions for the development of fungal decay. It is essential to ascertain that putty at the back of the glass extends down to and under the base, and there must be sufficient putty on the face of the glass and under the external glazing beads to ensure that water is totally excluded. If it is not practicable to remove a glazing bead during a reconnaissance survey to check that there is sufficient putty, the moisture content of the horizontal timber member below the glass should be checked with a moisture meter, particularly where a vertical member, such as the stile, jamb or mullion, is housed in the horizontal member. In the absence of putty under the glazing bead, it is not unusual to find that the moisture content in the horizontal members is in excess of 24 per cent and, if a bead is removed, silt and algal growths on the glass or putty will often be found. Early paint failures in horizontal members under external glazing beads usually indicate that the puttying is inadequate. In such cases it is usually necessary to recommend complete reglazing, using an oil mastic, and bedding the beads in the mastic. A rebate on the underside of the glazing bead, against the glass, is to be recommended, because this provides a 'sump' of mastic, which will guard against entry of moisture should the mastic at the top of the bead harden and shrink. Glazing beads should, of course, be mitred at the angles and should not project beyond the outer edges of the rebates for the glass.

Timber references

The Building Research Establishment has issued Digests and Information Papers dealing with the recognition and cure of timber defects and the precautions to be observed to minimise the risk of decay in modern joinery, and these include:

Digest 201 'Wood preservatives : pretreatment application methods'.
Digest 261 'Painting woodwork'.
Digest 262 'Selection of windows by performance'.
Digest 296 'Timbers : their natural durability and resistance to preservative treatment'.
Digest 304 'Preventing decay in external joinery'.
Information Paper IP 10 / 80 'Avoiding joinery decay by design'.
Information Paper IP 2 / 81 'The selection of doors and doorsets by performance'.
Information Paper IP 16 / 81 'The weatherstripping of windows and doors'.
Information Paper IP 21 / 81 'In-situ treatment for existing window joinery'.
Information Paper IP 7 / 83 'Window to wall jointing'.

Basically, the recommendations are that timber for joinery manufacture shall be adequately dry, suitably treated with wood preservatives, adequately protected in transit and during storage on site, properly primed with a good grade of primer, that attention should be given at the design stage to ensure that water will not be led into joints subsequently, and that if the unit is cut on site, re-priming is carried out immediately.

Door and window furniture

As to door furniture, it should be recorded whether the locks are rim or mortice locks, and whether there are keys for them. If keys are missing, apart from the material cost of a new lock, its installation cost must also be considered. The condition of spindles should be checked by shutting and opening the door, and turning the handles to ascertain that the latches are in good working order and that keys fit properly. Missing escutcheons and damaged or missing finger plates should be recorded because matching replacements may be surprisingly costly, or unobtainable.

Window catches, casement stays, cockspur fasteners and the sash cords to double-hung sashes must be checked. Great care should be taken when undoing catches to double-hung sash windows in case the cords of the upper sash are broken. Every opening window, casement or fanlight should be opened to ascertain that it is not distorted or stuck up with paint. If there are shutters they must be opened unless they have been permanently screwed or nailed up, when that fact should, of course, be recorded. The first traces of fungal decay are often to be found in the linings to shutter boxes because these are of thin timber, and closer to the external walls than the thicker linings and architraves to the window openings themselves.

Decorations

Although it can usually be assumed that the prospective purchaser has noted the condition of the decorations, the surveyor should comment on them. Those who are regularly carrying out reconnaissance surveys will be only too well aware that new decorations can cover many defects. Another aspect is that so many owners do their own decorating today, and the quality of the work may leave much to be desired. There comes a time when the calling in of a professional becomes essential, and the removal of amateur efforts can add appreciably to the cost of re-decorating. It is as well to stress that although the decorations are 'clean', once the existing furniture and pictures are removed, they may well no longer be acceptable.

Paintwork

The condition of the external paintwork must always be commented upon in the survey report, and whether the work has been done professionally. When owners do their own maintenance the preparatory work is often inadequate, and actual defects, for example, patches of wet rot, may have been painted over. Sooner or later, a new owner is faced with having to strip the old paint and this adds appreciably to the cost of re-painting. If the surveyor has to advise stripping off the paint, the danger arising from burning off should be stressed. Old box frames present a particular hazard if there are traces of old creepers adhering to the walls, which may also have grown behind the box frames. It is the responsibility of the builder if the carelessness of his men results in a property being burned down, but not all firms are adequately insured to meet substantial claims. The inclusion of very strict clauses relating to the use of blow lamps to burn off paint and to repair lead work is recommended and it must be made abundantly clear that responsibility rests with the builder. Nevertheless, it is advisable to insist that the use of blow lamps shall cease at a reasonable time before any breaks for meals and at the end of the day, before the men leave the site. Moreover, an efficient and fully charged fire extinguisher should always be available when a blow lamp is in use.

Fixtures and fittings

The surveyor should ascertain what fixtures are included in the sale because removal of these may well put the new owner to considerable expense in making out cornices, picture and dado rails and skirtings stopped at a wall fitting that will be removed. The condition of fixtures and fittings to be left should be inspected as thoroughly as the more important structural features described in preceding chapters. If fixtures and fittings are of timber, being 'joinery' items, they are likely to be made up of thin timbers, which may contain considerable quantities of sapwood and may therefore be susceptible to furniture beetle attack or, if of hardwood, to *Lyctus* infestation. The surveyor should not be misled by fine dust under drawer runners to guides, produced by friction of the drawer sides. This often has the texture of *Lyctus* frass but, when examined with a pocket lens, will be seen to be quite distinct.

8 Services

Within a structural survey report, internal plumbing, sanitary fittings, external drainage, electrical and gas supplies and central heating should generally be classified under the heading of 'services'. In a detailed structural survey, these services will not be dealt with in isolation in the course of the room-by-room investigation. In smaller properties, however, it is convenient to record what services are present, although, again, the notes should be brought together under a separate heading within the survey report.

Water services

The provision of water is controlled by statutory regulations and the bye-laws produced by respective Water Authorities. They are based on the prevention of waste and contamination of water and the provision of an adequate and wholesome supply. Reference to the requirements of the local Water Authority will be the basis of establishing the adequacy of the storage and supply to a building.

The surveyor's attention is drawn to *BRE Digest 98* 'Durability of metals in natural waters', which discusses the properties of water that affect the durability of plumbing installations. In this Digest it is pointed out that mineral impurities such as chlorides and sulphates, together with dissolved oxygen and carbon dioxide, influence the corrosive behaviour of water. The effect of these impurities, including oxygen and carbon dioxide, is discussed in relation to the different materials used for storing and distributing water to sanitary appliances and central heating systems in buildings. This Digest is of such practical importance to the surveyor that no precis would be adequate, as will be appreciated from the summary of its scope that appears below:

"It deals with—the nature and concentration of impurities in water hardness—types of scale formed—prediction of scale formation—resistance to corrosion of commonly used metals—effect of scale formation on corrosion resistance—hot water systems—

the risks of using a mixture of metals—protective measures—loss of zinc from brass components—heating installations—means of avoiding corrosion and 'air' locking—protection of pipes against their external environment''.

In the bathroom and WC the surveyor will note the sanitary fittings, for example, the bath and hand basin, their size and condition. WCs should be checked for leaks and systems flushed. The pipe runs should be carefully examined because in older properties they may not conform to the recommendations of the Water Authority. The surveyor should note that, in some areas, lead or galvanised steel barrel may be positively unsuitable because of the nature of the water supply. The most common failing with pipework is to find a mixture of galvanised steel and copper which, if in the hot supplies, can result in electrolytic action between the two metals, thus causing corrosion. This is often the case where tappings from galvanised cylinders are in copper, and brass unions have not been used. Old lead pipes, particularly those buried in the wall, are a frequent source of minor leaks from small holes. Plumbing in old properties is often such a tangle of pipes that it is very difficult to detect leaks. In these cases it may be advisable to recommend complete renewal. It is important to check whether domestic hot water is supplied by an indirect cylinder or direct from a boiler. In the latter case the pipes and boiler in a hard water district are liable to fur up after a few years, when they need to be cleaned chemically by a specialist or renewed *in toto*.

The hot water installation, including the central heating systems, should be examined for condition and serviceability; insulation and pipe jointing are obvious points to note and so too are the controls (thermostats, safety valves and time control mechanisms). Numbers and positions of radiators, boilers and immersion heaters should be recorded. It is difficult sometimes to establish inefficiency in a hot water installation apart from the boiler and its ancillary equipment and the builder's work in the flue, but an assessment should be attempted even if it means a specialist investigation at a later date.

If the client is buying a property with central heating for the first time he should be made acquainted with operating costs, as these can come as something of a shock to the inexperienced, particularly if a high-cost fuel is used and the building is poorly insulated.

Drainage

There are few, if any, moving parts in a drainage installation and even these can be kept working efficiently by routine cleaning. Tracing and testing of the drains obviously forms part of the external investigations, but unless the firm carrying out the survey uses a member of the staff to investigate, who is also responsible for checking on the internal plumbing, sanitary fittings and connection to the drains, the work will normally be deferred until the interior survey has been completed.

Many firms prefer to leave the testing of drains to specialists, and there is much to be said for this course, but as specialist services add to the total cost, the surveyor should first satisfy himself that the condition of the property justifies pursuing negotiations. In a detailed structural survey, the drains should in any event be dealt with entirely separately, because the necessity for such a survey in many cases presupposes adaptation of the building for some entirely different purpose, when the existing drains and other services are likely to be inadequate.

When a large property is being sold off in lots and the surveyor's client is interested only in one, warning should be given where the drainage of all the lots is to one cesspool or septic tank, that subsequent re-development of the other lots may well overload the existing sewage disposal arrangements.

Testing of drains, whether done by the firm carrying out the reconnaissance survey or by specialists, means additional expense. The client however, must be left in no doubt of the desirability of a proper test when all the surveyor has done is to trace the layout of drains, located manholes and seen that they appear to be functional. These limited investigations are essential because they may bring to light matters sufficient in themselves to necessitate advising against pursuing negotiations. For example, in a farm house in Somerset equipped with 'modern sanitation', tracing of the drains revealed that the soil drain discharged into the Local Authority's storm water sewer. In this case the prospective purchaser was contemplating the addition of a second bathroom and a ground-floor cloakroom/WC, which would, of course, have required planning permission. The council would almost certainly have insisted on the installation of a proper sewage disposal scheme, which would have put the new owner to expenditure out of all proportion to the value of the property.

Normally, sewage disposal is by one or other of three means: to a public sewer, to a cesspool, or to a septic tank. It is the duty of the surveyor carrying out a reconnaissance survey to establish which of these three alternatives exists. Even where the system is mains drainage, faults may nevertheless exist. Not infrequently in country properties, facilities have been added to without the permission of the local Environmental Health Officer. Connections to the main drain may have been made with Y-junctions, without manholes, interceptor chambers or rodding eyes, so that in the event of blockage it is impossible to rod some lengths of drain. Additional hand basins may have been added, discharging into the storm water drains which, in turn, discharge into soakaways in breach of bye-laws and the current Building Regulations. If the system discharges into a cesspool, there is no certainty that this is ideally located so that it can be emptied periodically, with the result that the effluent from the cesspool may, for example, be causing pollution of streams. Even with a properly designed septic tank, it is still necessary to ascertain that it is of sufficient capacity, and that the arrangements for disposal of the effluent from the tank, however pure, will not ultimately cause flooding of a neighbour's land.

Modernising of existing systems and correcting faults in design and lay-out of both soil and storm water drains can prove very costly, and it must be made perfectly clear in the survey report how far investigations have been taken, stressing the advisability of having the installation fully investigated and tested. Even where there is mains drainage there is still room for complications. For example, with small suburban estate properties, the main soil drain is often the joint responsibility of several separate owners. In these circumstances, it is imperative to stress in the report that the rights and liabilities of the purchaser in respect of the drains must be investigated by the purchaser's solicitor, in order to establish the obligations of any one owner.

Electrical installations

It is most important that the installation be inspected and in some cases tested, when either newly installed or an existing system is being considered. There are several activities involved, for example a visual inspection could show whether or not the installation and associated equipment comply with the appropriate standards which are listed in the IEE

(Institution of Electrical Engineers) Regulations. This compliance must also extend to the components and the wiring techniques that have been adopted. At the same time it should be obvious if the installation has been damaged in any way, for example by subsequent building work.

The tests which are to be carried out are usually undertaken by a specialist and the order of testing is very important for the test itself may be a cause of danger if the installation is faulty, especially so if any protective device fails to function. In addition, some tests rely on other parts of the installation and this depends therefore on the sequence. The testing sequence normally undertaken is:

(a) ring circuit continuity;
(b) protective conductor continuity,
(c) measurement of earth electrode resistance;
(d) measurement of insulation resistance;
(e) protective efficiency of barriers and enclosures;
(f) measurement of resistance of non-conducting floors and walls;
(g) certification of polarity;
(h) measurement of the earth fault loop impedance;
(i) testing earth leakage circuit breakers.

Equipment and knowledge for carrying out testing to this standard is not usual equipment for surveyors. However, it is important to know what instructions are appropriate to be given to an engineer and it is essential to understand the highly technical report that may result from such a test. When a system is newly installed, there should be evidence that it is efficient and safe. This is achieved by the contractor issuing a 'Completion Certificate' to which will be attached an 'Inspection Certificate' indicating that the work has been tested and found efficient and that the installation does comply with the IEE Regulations. The 'Completion Certificate' should be signed by a qualified person.

Gas

Gas supply in the UK is administered by the British Gas Corporation through twelve gas regions. Installation of gas services can only be carried out by the gas region's engineers or members of the Confederation of Registered Gas Installers. When inspecting a property with a gas installation, the surveyor should advise, as a matter of routine safety, of the need to inspect and test the service shortly after purchase.

If the house is old, there is a likelihood that the old supply pipes may still be in use and these pipes may be leaking since modern natural gas is supplied at higher pressures than the old town gas. In addition, where gas lighting or old gas fires have been removed, the original pipes may have been capped off and left hidden within the building. Structural movement caused by renovation work or traffic vibrations may fracture these live pipes, causing a slow leakage.

Where gas fires, boilers and other appliances are installed, adequate ventilation must be provided for their safe operation. Typically for safe running, a single air brick of 225 x 225 mm (9 x 9 inches), incorporated into the external wall, should suffice. If there is no method of permanent ventilation, then the surveyor must advise his client of the possibility of toxic poisoning and suffocation when the appliance is in operation.

It is popular in some houses which do not have central heating to make use of wall-mounted gas heaters. These usually are vented to the outside via balanced flues. The surveyor must check that these flues are not positioned directly under openable windows. If this happens then there is a danger of waste gases finding their way back into the dwelling.

9 Report Writing

Surveyors who are capable of observing accurately, occasionally fall down when it comes to compiling a report of their observations, including the interpretation of their significance. As far as the client is concerned, it is a lucid report that matters since he has retained the surveyor to observe and diagnose what can be found out about the property, short of taking it apart, that he cannot discover for himself as a layman. In some cases there is little point in giving the dimensions of rooms or in describing the condition of the decorations in detail, for the client should be given credit for having read the estate agent's 'particulars' and for having seen the obvious, otherwise he would not seriously be considering purchase.

When a surveyor has been instructed, the purchase will probably be subject to a satisfactory report being received from the surveyor, the title and other legal aspects being the responsibility of his solicitor to discover. Therefore, what the client wants to know from the inspection is whether the property is basically sound and, if not, how much will it cost to carry out necessary remedial work. The surveyor must always bear in mind that ultimately, his survey report will be looked upon in monetary terms by the client, that is, how much to put it right?

The written word

It is rarely possible for a surveyor to draft a preliminary report that he can subsequently work up into a refined form, and consequently the initial draft should be carefully considered from the beginning. What is required is a lucid description of the condition of the property and the significance of its defects, expressed in plain, simple English. There is, of course, every reason for a surveyor to endeavour to improve his writing ability, which is something that does not always come easily. Some works of reference are useful aids to clarify expression, and the correct use of English. Among these, Fowler's *A Dictionary of Modern English Usage* is particularly helpful. Many may not agree with Fowler's punctuation and rigid rules, but a study of his book should make for consistency in the written word for those without a natural gift for writing. Further, although one may not be given to malapropisms, a good dictionary should be on every surveyor's bookshelves. Another book which can be recommended is Gower's *The Complete Plain Words*, now revised by Sidney Greenbaum and Janet Whitcut and published by HMSO.

When writing a report, certain rules must be observed. First and foremost ask the question: for whom is the report intended? The detailed structural survey, although probably commissioned by a layman, will primarily be studied by a client's architect or structural engineer, whereas a condition survey report may be intended for the lay client. This type of client may have a horror of dry rot or woodworm, probably without knowing why, and almost certainly without appreciating that the finding of either may be no reason for abandoning the idea of purchasing the selected property. Most technical terms and scientific names will be meaningless to him and so they should be avoided or explained. In certain circumstances it may be imperative to use technical terms, which should then be simply defined.

Some reports are written as letters, commencing and ending conventionally, but this format is not to be recommended. A report is an important document and should have a definite title, for example, 'Report on The Gables'. If the property is a large one and the first report covers no more than initial preliminary inspection, the title should disclose the fact: 'A preliminary (or interim) report on a reconnaissance survey of The Gables'. Paragraphs should be numbered and each separate section should have a heading, so that vital matters can be clearly indicated in the covering letter sent with the report, or referred to in an appropriate section.

Limitations

It is essential for the surveyor to make it clear to the client how far his investigations have gone and whether it is advisable to take matters further by opening up, testing of services and the like, before a final decision to purchase is made. In surveying a fully furnished house the surveyor may be unable to take up carpets, lift parquet flooring or tongued and grooved boarding, move furniture, etc., without authority and possible risk of damage. If this is the position, the surveyor must state precisely in his report what he has and has not been able to do, for example, 'room close-carpeted and flooring not inspected' or 'room partially carpeted and margins or flooring only inspected'. Such limitations as fitted carpets and linoleum obviously make it difficult for the surveyor, but it is no justification for introducing large numbers of escape clauses into the report that make the document virtually worthless to the client. Moreover, as is pointed out in chapter 10 on Legal Aspects, escape clauses will not exonerate the surveyor if he has been negligent.

It is the surveyor's duty to assess the significance of what he has observed and if necessary to underline the inevitable limitations of any deductions. For example, it is of no assistance to the client if his surveyor merely states that he found no traces of woodworm but can give no guarantee that none is present. If he can say that he has made a thorough search in the cupboard under the stairs, the plywood backboards for the electricity intake and gas meters, the margins of floors in any rooms not close-carpeted, the frame of the access hatch and the boarded areas in the roof void, and that he saw no signs of recent or active infestation in those places nor in the furniture, and further that there were no articles in the house likely to encourage the borer (for example, wickerwork waste paper and linen baskets), then this amount of information will be of considerable value to the client. Of course it cannot be backed up with an absolute guarantee that there is no infestation, but it does indicate the extent of the investigation.

On such negative evidence of infestation in a house more than ten to fifteen years old (that is, a house of sufficient age for there to have been time for initial infestation and subsequent re-infestation and provided the amount of sapwood in softwood carcassing timbers and flooring is not excessive), one may be justified in stating in a report that in the surveyor's opinion there is little risk of there being any serious furniture beetle infestation in the property. He would add, however, that it would be advisable to treat valuable pieces of furniture prophylactically, but he would give the same advice to anyone moving into a flat in a modern block that contained little or no timber, because even a small amount of attack in such furniture has a disproportionate effect on values.

Similar deductions are permissible in regard to dry rot; for instance, if as a result of visual inspection, supplemented by probing timber in vulnerable positions, none of the symptoms are found, a surveyor is justified in being reassuring, provided there are no features calculated to induce future outbreaks. These features could be the absence of through and cross-ventilation under suspended ground-floor floors, ground-floor rooms with timber floors covered with thermoplastic tiles, complex or inadequate rainwater disposal arrangements, perimeter walls so tight to boundaries that proper maintenance of gutters is not practicable, dense vegetation growing against external walls likely to obstruct air bricks, and so on.

Setting out

1. *Instructions*

Give the authority for carrying out the survey, that is, a letter or telephone call—the surveyor must confirm all his instructions in writing together with his conditions of engagement to the client, before carrying out the survey. Any special terms of reference should be set out, for example, the client's requirements or any information the surveyor is given regarding the property. If the information he is given is extensive it is preferable to make a separate section: 'Relevant Information'. The outcome of the surveyor's own researches will also be recorded in this section.

2. *Scope of investigations*

The date or dates when the survey was carried out and the weather conditions should be stated. Also, whether any building operatives or specialist contractors were in attendance with the surveyor. The scope of the enquiries should be recorded in this section, for example, inspection of deposited plans such as the Development Plan for the area or the Geological Survey map, etc.

3. *Relevant information*

The surveyor should distinguish between information supplied by the client or solicitor, and information found out for himself, giving the sources of the information, for example, the various departments of the Local Authority. The sort of information the surveyor may include can cover such things as the Highway Authority's proposals for road-widening projects or construction of new roads that may adversely affect the property (or be of positive advantage), and/or relevant information from the Geological Survey maps. Other matters to be covered could include the following:

(a) Structural alterations or additions.
(b) Any record of flooding and, if so, whether flood cover can be obtained at normal or heavily weighted insurance rates.
(c) Any previous history of dry rot, beetle infestation or underpinning.
(d) The type and proximity of trees from the point of view of potential hazard and whether any trees are the subject of a Tree Preservation Order.

4. *Site observations*

Although there is no ideal, these can be set out in the following framework:

(a) generally;
(b) externally North elevation, West elevation, South elevation, East elevation;
(c) roof coverings;
(d) rainwater disposal arrangements;
(e) outbuildings;
(f) grounds, boundaries and fences;
(g) internally, the roof voids, rooms floor by floor;
(h) services, heating, electrical services, water, plumbing and drains.

It is possible to draft check lists carrying out inspections but, as with all *aide-mémoires*, there are dangers in relying exclusively on a predrafted list; however, it is useful to have a routine procedure and this may be helpful to those beginning their surveying career. On the check list, provision should be made for comment, although this could be recorded elsewhere and a simple coding introduced on the list. For example, the section illustrated could be used for the inspection and only when the coding is 'defective' would a further note be prepared. External works could be recorded in this form, see figure 22. Similar classifications could be applied to the elements of structure, finishings and services, and elaborated when necessary after experience.

Certain aspects of the presentation leave room for personal preferences. Some surveyors prefer to bring together all the observations on, for example, joinery, although the observations will have been made room by room. Similarly, light and power points and radiators can be listed within schedules rather than being recorded in the descriptions of the individual rooms. This grouping of repetitive data makes for brevity but involves more work when drafting the report as the information has to be separately abstracted from the Site Notes. It is usually an advantage to a client to have schedules of the electrical services and radiators as he will be more readily able to decide where he requires additions to these services. Grouping of information relating to joinery, both windows and doors, and to flooring, is really only satisfactory in smaller properties where these items are standardised throughout the house. In larger and older properties, there is likely to be so much variation in the items themselves, and their condition, that it may become necessary to list defects in rooms where they exist. It is, however, helpful to include an introductory paragraph on the

EXTERNAL WORKS.								
		CONDITION.			PRIORITY.			
	TYPE	GOOD	ADEQUATE	DEFECTIVE	LOW	MEDIUM	HIGH	COMMENTS.
PERIMETER/BOUNDARY.								
HARD LANDSCAPE.								
LANDSCAPING.								
TREES.								
DRIVE.								
PATHS.								
GULLIES.								
DRAINAGE.								
MANHOLES.								

Figure 22   Check list

type and condition of windows, doors and flooring, at the beginning of the section dealing with the room-by-room condition of the property. This will reduce repetitive information in the descriptions of each room.

5. *Dilapidations*

These are essential works required to put the property in a good state of repair and which are to be priced. Even in a reasonably well-maintained property, quite large amounts of money could be required to restore the property to a good state of repair, and such sums may well be beyond the means of a prospective purchaser who has budgeted only for the purchase price. Itemising the cost of dilapidations in a really old property is all the more important if it is known that the client has limited resources. This applies particularly to dilapidated properties in areas that are changing character and becoming fashionable. Before accepting instructions the surveyor should ascertain the client's intentions regarding modernisation and approximately how much he intends to spend on the property. For an approximate budget figure, little more than walking round the property coupled with taking a few measurements will likely be sufficient for an experienced surveyor to suggest an amount that is appropriate for repairs, rehabilitation and any extension.

Dilapidations with regard to landlord and tenant agreements require a different approach to that undertaken for prospective house purchasers. The *Royal Institution of Chartered Surveyors' guidance note* 'Dilapidations' and Professor Ivor Seeley's book *Building Surveys, Reports and Dilapidations* are excellent sources of reference for those surveyors who undertake such work.

6. *The client's special requirements*

These can be the provision of an additional bathroom or even major structural alterations. It is important to indicate what is feasible, putting forward alternative suggestions that would meet the requirements, but possibly at less cost. Each requirement should be spot-priced as far as is practicable because laymen often have little idea of how much an apparently simple requirement will cost. For example, few laymen appreciate that an external toilet could involve extensive drainage costs.

7. *Conclusions*

This section of the report is of vital importance to the client, and it is here that the surveyor must use his professional skill and judgement in explaining the significance of the facts recorded in the body of the report. If he includes numerous escape clauses or repeatedly suggests calling in commercial firms to advise on matters that should be within the province of a competent surveyor, the report will be of little value to the client and the brief may not have been fulfilled. Continual resort to such evasion may lead to a decline in standards and an inevitable loss of public confidence in the surveying profession. One must, however, bear in mind the attitude of litigation-hungry consumers who often look upon the survey as a form of insurance should 'anything' go wrong. In this regard it is important to make it clear what are 'observations' and what are 'deductions', that is, the surveyor's interpretation of the significance of what he has observed. It may be necessary or desirable to comment on observations in a particular section of the report but these comments or deductions should be brought together under Conclusions. In this way there is no possibility of the client overlooking their significance. It is advisable to draw attention to the Conclusions in the letter accompanying the report.

Report writing synopsis

The purpose of a report is then to assist someone to come to decisions that are reasoned and these can be of different degrees of importance to different clients. Therefore all relevant matters should be included and their relative importance in the context of the brief made clear. Reports will vary in length, format and style, but the structure can be similar and include:

(a) title;
(b) for whom;
(c) by whom;
(d) introduction and arrangement;
(e) sections following the above;
(f) conclusions (summary of critical points, recommendations, etc.);
(g) appendices (drawings, etc.);
(h) references (documentary sources, etc.).

The style of writing is influenced by the purpose of the report and the client, but it should be direct and simple with words chosen that have common usage. Sentences and paragraphs should be short and punctuation used to improve clarity. When more than one surveyor is involved, consistency in style and presentation may be difficult to achieve, but nevertheless it should be an objective and may be imposed by editing. A report is usually read when the author is not present, although occasionally a report may be presented. This fact is sometimes not appreciated and consequently, ambiguous or emotive words and phrases should be carefully avoided.

When the draft is completed, editing must commence and this should be done to improve clarity and reduce the length by omitting all that is ambiguous and not absolutely essential. Abbreviations and statements of the obvious can reduce the value of the report to the reader. Grammatical checking together with the checking of any calculations should also be done and then the report should be reviewed from one point of view—that is, has the original brief been completely fulfilled, even though possibly refined and clarified by the surveyor. A typical reconnaissance survey report is shown in appendix C.

10 Legal Aspects

Introduction

He who is his own lawyer has a fool for a client. This chapter is not written in disregard of that prudent aphorism. Its object is merely to provide surveyors with some background knowledge on points of law that often affect them or their clients, to alert them to points to guard against. It is not intended to be a substitute for proper legal advice when specific problems arise. The surveyor should then consult his solicitor or advise the client to consult his. If advice on the law is given, then unless the surveyor is sure of his ground he should check it with his own solicitor and confirm or correct it in writing to the client. The surveyor should in each such case consider giving the advice with an express disclaimer of responsibility. Coming from a lawyer, advice to use lawyers whenever possible may be viewed with scepticism. But it is given secure in the knowledge that it will continue to be widely disregarded and thus much business for lawyers will continue to be generated. Neglected legal problems almost always turn into bigger and probably litigious legal problems. Yet litigation is extremely expensive and unless the case is fairly simple and straightforward the costs often make it unworthwhile unless the sum involved is well over about £25 000 (at present money values), and even then the relationship of costs to the amount at stake and the prospects of success have to be carefully watched throughout. Because of the unjust rules of legal costs taxation, the successful litigant only recovers about a half to three-quarters of his actual costs of litigation from the other party. The rest has to come out of the damages. Hence the importance of insurance of all possible liabilities and losses, including fees recovery.

Throughout this chapter the male gender is used because 'he' is shorter than 'she' and shorter still than 'they'. In the main the law is stated as at the 31st December 1986.

Legal liability of the surveyor

There are no special rules governing the liability of a surveyor. His legal position is governed by exactly the same rules as those which apply to any other citizen. The object of the first part of this chapter is to consider those which affect him most especially.

When he accepts instructions, of whatever nature, from a client he thereby enters into a contract with that client. Even if there is nothing in writing it is an implied term of the contract that he will carry out his instructions with reasonable care and skill. This simply means that he must not be negligent. In theory he and the client could agree to increase or limit his liability but in practice this does not occur. No reasonable client can expect a higher standard than reasonable care, and if a surveyor were dishonest the client would always have his remedies in tort. No reputable surveyor would want to be anything less than careful, and he would hardly increase his practice if he insisted on excluding his liability before accepting instructions.

What constitutes reasonable care depends on all the circumstances. Previously decided cases are useful guides, but it is always important to consider all the relevant facts for there may be important points of distinction between the instant case and previous decisions. A particularly expert surveyor, who holds himself out as such, must show more than usual expertise in his special field in order to discharge his duty of care. The client engages him on account of his special skill, and it is implicit in the contract that he will display such skill. On the other hand, the standard of reasonable care is a minimum one, and if a surveyor accepts instructions it is no defence to an action for professional negligence to say that he was very inexperienced, or the instructions were outside his province. The inexperienced must be extra careful on account of their lack of practical knowledge, and any man who accepts instructions should either ensure that the work is within his province or else learn all about it before doing it.

A surveyor may be a member of a professional team of experts. He may discharge his duty of care to his client by relying on another specialist where appropriate but the extent to which he has done so should be made clear in writing to the client. In such circumstances it is also important before embarking on the commission to clarify in writing who employs whom so that problems of hierarchy of authority are reduced and the paymaster of each expert is clearly identified.

Insurance companies often insist that reports on surveys shall include a clause to the effect that although no timber decay or beetle infestation has been found no guarantee can be given that there is none. Though sometimes called 'exclusion clauses' these are not really such. They do not exclude the surveyor's liability for the work which he has done. They simply indicate the limits of that work. It is important for the client to know how much reliance he can place on the report, and whether a more extensive inspection is necessary to investigate some aspect that the surveyor was unable to pursue, either because he did not have the necessary technical knowledge, or because facilities were not available to him at the time of his survey. For example, a modern building may incorporate new building techniques with which the surveyor is unfamiliar and of whose safety and durability he knows nothing. In such cases he should either say so, or acquire the necessary knowledge before reporting. Alternatively, in an occupied house it may not be possible to inspect all floors because of floor coverings. Again, the surveyor should say so, and if, having regard to the state of the parts which he can see, there is any risk of timber decay or beetle attack in the concealed parts, he should point this out. He must remember that the client is a layman and it is not enough to tell him of the symptoms from which conclusions can be drawn. The surveyor is employed to draw those conclusions and tell the client of them. Thus not only should signs of dampness or lack of ventilation be reported; the surveyor must also warn that these may have produced timber decay in the areas affected.

The duty of care under the contract is owed to the client. In many cases a solicitor will also be involved. *Vis-à-vis* the surveyor he may be the lay client's agent, or the lay client may be his, or they may both be principals. Generally speaking only the principal to a contract can sue or be sued on it. The latter aspect is important when a surveyor has to consider whom to sue for his unpaid fees, and this is one of the reasons why it is important from his point of view always to have his instructions clearly expressed in writing, for then there can be no argument afterwards.

It sometimes happens that the client wishes to show the surveyor's report to other people for them to rely on it, as where purchase moneys must come out of a trust fund and the intending buyer who gets the report needs to show it to the trustees to get them to make the necessary money available. If the surveyor has no idea that this may be done he owes no duty to those other people, and if they rely on his report and because of his negligence suffer damage, they have no right of redress from him. But if he knows or ought to know that his report will or may be used for this sort of purpose he owes a duty of care to those people whom he should reasonably foresee will rely on his report. On the basis of the same principle a surveyor may be liable to those to whom he gives professional advice informally. Thus if a friend or relative seeks his professional advice in circumstances where he should realise that this advice will or may be relied on by that person, he owes a duty to him to take reasonable care, and if he gives negligent advice he will be liable. But not every expression of his professional opinion carries with it this legal duty of care. If a surveyor expresses his opinion on a matter within his professional sphere on some social occasion, he owes no legal duty of care to his hearers. But if one of them seeks his technical advice on some problem of his own, in circumstances where it is plain that that advice is going to be relied upon, a legal duty of care is owed, and if he is negligent in giving that advice he will be liable (*Hedley Byrne & Co. Ltd* v. *Heller & Partners Ltd* [1964] A.C. 465). An important recent example of these principles is provided by *Yianni* v. *Edwin Evans & Sons* (1982) Q.B. 438. A building society engaged a firm of surveyors and valuers to value a house for a prospective mortgage. The prospective purchasers, the plaintiffs, had to pay for the survey. They were advised by the building society to have their own survey done but did not. The surveyors assessed the house as suitable for maximum lending, negligently failing to discover defects whose repair cost exceeded the value of the house. The house was duly bought by the plaintiffs relying on the fact of the surveyors' valuation. The surveyors had realised that they would do so and accordingly were held liable to them.

Apart from these duties owed in contract or situations analogous to contract a surveyor owes the ordinary duties in tort. In carrying out his survey he must not go on to the property of others without their consent or he will be liable in trespass. Where a house is in multiple occupancy, in the case of a lease, or where it is necessary to go on to neighbouring land in order to make a proper inspection of the property under survey, the surveyor should make sure that he has permission from each of the parties concerned to go on their property, or else that he has a suitable indemnity from his client. In making his report a surveyor may sometimes have to make critical statements about others. Provided that they are true statements of fact they are not defamatory. But the truth is often difficult to prove and unsuccessfully to plead that an alleged libel is true only aggravates it. The expression of an opinion which is fair comment on a matter of public interest, without malice, is not actionable. Furthermore, a report made to a client is protected by qualified privilege, which means that a defamatory statement in it is not actionable unless it was made with actual malice.

In order to preserve that privilege, and to avoid possible liability to third parties, it is wise to include in a report, or a covering letter with it, a statement to the effect that the report is only for the use of the client and no liability or responsibility can be accepted if it is shown to or relied on by others.

Liability of the surveyor for the acts of others

A surveyor is generally only liable for the acts of his own direct employees. Where a number of experts are engaged on a project, clarification of their spheres of responsibility is needed for avoiding demarcation disputes and to show how far they can rely on each other, as stated on page 132 above. But unless a surveyor employs someone else to perform his own contractual obligations he will not generally be liable for the torts of independent contractors engaged by him.

Two particular cases warrant specific mention.

(i) *Calling in of specialist firms: surveyor's liability*

If a surveyor recommends an incompetent firm without having taken reasonable steps to check its competence and the client suffers damage as a result, the surveyor is liable, not vicariously for its incompetence, but for his own negligence in failing to discover this beforehand. But if he has recommended a firm which he reasonably believes to be competent, and it is in fact guilty of negligence on this occasion, he is not liable. Thus if the firm does its work badly, recommends a treatment where none is necessary, or in performing necessary work includes unnecessary or extravagant items, the surveyor is not liable. This is because his liability for the acts or defaults of others is confined to his direct employees and does not extend to independent contractors.

(ii) *Liability of a builder who causes a fire and of surveyors who recommend work involving a fire hazard*

The builder is liable if he has been negligent. Since the dangers of using fire in the vicinity of combustible material are obvious and great, he is most unlikely to have been sufficiently careful if a fire does break out in the house from his work. He is liable not only to the home owner but to the owners of adjoining properties to which the fire spreads.

The house owner who employed the builder is liable for his negligence in this respect even though the builder is an independent contractor *(Balfour v. Barty-King* [1957] 1 QB 496).

The surveyor who advises burning off, etc., must take reasonable care to see that it can and will be safely done. The amount of supervision required of him depends on the competence and experience of the builder. The greater these are, the more the surveyor is entitled to leave the job to him. It is wise for the surveyor to state in writing the precautions to be taken. If he wishes to be sure of placing all possible liability on the builder's shoulders, completely exculpating himself, he must obtain an indemnity from the builder. Clear language is needed to avoid liability for negligence should the surveyor in fact have been negligent.

Similar principles apply to other extra-hazardous activities.

Professional Indemnity Insurance

No human being can never be negligent or the victim of unfounded allegations of legal liability, but his worries will be reduced if he is adequately and safely insured.

Such cover should be effected by a surveyor through substantial brokers specialising in such insurance who will advise him properly about it and who will have influence with insurers if they question whether a particular claim is covered by the policy. It is important for the surveyor to liaise closely with his brokers about all insurance aspects, preferably in writing to avoid disputes about what has been said.

Four points are especially important.

Upon proposing for new insurance and at each renewal it is the duty of the surveyor to disclose to insurers all facts which a prudent insurer would consider material in deciding whether to accept the insurance or at what premium. This duty is draconian. It may be extended or modified by a proposal or renewal form but it is not necessarily thus modified. If it is not then accurate answers to questions in the form will not alone necessarily discharge the duty. The position in each case is a question of fact depending on its individual circumstances. If the duty is not performed, or if there are inaccurate answers in the proposal form, then in general insurers will be entitled to avoid liability for any claims under the policy, even if there is no connection between a particular claim and a fact wrongly undisclosed or mis-stated and even though the surveyor has in no way been fraudulent. It is therefore highly desirable to have the help of a suitable broker in completing a proposal or renewal form and to record in writing to the broker at the time the fact and extent of such help and the surveyor's dependence upon it. There should be reliable internal office procedures to ensure that all relevant information is obtained from partners and staff.

Insurers can refuse to deal with a claim on the policy which is not presented in accordance with its requirements as to such matters as time ('immediate notice') or form ('in writing'). The surveyor should get the guidance of his brokers on such aspects and ensure that his partners and staff are familiar with them.

Damages for all types of claim have increased enormously over the last ten years or so, and will go on rising roughly in proportion with inflation. It is therefore essential to keep levels of insurance cover adequate.

When a surveyor retires or dies it is important that insurance cover is continued since claims may still be made in respect of his alleged negligence or other breaches of duty whilst he was still in practice. Such claims may be made for as long as the law of limitation of actions allows (see below). In general, time does not run until damage occurs which may be many years after the events from which the claim arises. Accordingly proper professional advice should be sought as to the length for which insurance cover should be maintained. This will be at a substantially reducing premium as claim-free years go by.

Since the 1st January 1986, members of the Royal Institution of Chartered Surveyors have been required to have professional indemnity insurance. Details on types of policy available to members and other relevant information may be obtained from RICS Insurance Services Ltd, Plantation House, 31–35 Fenchurch Street, London, EC3M 3DX.

Public Health and Environmental Requirements

(i) *Public Health Requirements*

Modern legislation contains a number of detailed requirements about the healthiness, amenities and safety of buildings. Many of these are dealt with elsewhere in this book, but the provisions are too complex and lengthy for a useful summary of them here. For that purpose the surveyor needs to consult specialist publications on these topics, attend courses upon them, or consult his own lawyers.

A few words may however usefully be said about some aspects of environmental protection (apart from planning and pollution, substantial topics in their own right).

(ii) *Preservation Orders on Buildings*

There are three types of 'preserved' buildings:ancient monuments, buildings of special architectural or historic interest, and buildings in conservation areas.

(a) *Ancient monuments* are buildings and structures listed as such under the Ancient Monuments and Archaeological Areas Act 1979. Their guardianship and maintenance are the responsibility of the Department of the Environment, the Historic Buildings and Monuments Commission, and Local Authorities. Listed ancient monuments may not be demolished, damaged, added to or altered without the written consent of the Secretary of State. Compensation and grants are payable in various cases.
(b) *Buildings of special architectural or historic interest* are buildings of such interest which on that account have been made the subject of a preservation notice by a Local Authority or have been included in a list of such buildings by the Department of the Environment. The provisions which apply to them are similar to those for ancient monuments.
(c) *Conservation areas* are so designated by local authorites and the Secretary of State. Buildings in them are subject to similar provisions to those just described. Trees in them are subject to similar restrictions as if covered by a tree preservation order.

(iii) *Tree Preservation Orders*

These are made by Local Authorities and prohibit the felling, topping or lopping of preserved trees without the Authority's consent unless such steps are necessary to abate a nuisance or because the tree is dying, dead or dangerous.

Sundry aspects of property ownership

(i) *Liability of property owners to others*

(a) *To passers-by on the highway.* The building occupier may be liable in nuisance or in negligence to passers-by on the adjacent highway for damage caused to them because of defects in his property. Who is the 'occupier' for these purposes is a question of some complexity, particularly in the case of demised premises, and cannot be dealt with fully in a work of this kind. The owner out of occupation may in some circumstances be liable. Liability in nuisance in the circumstances referred to above is strict. Thus it seems that it is no defence that the defendant did not know, nor should have known, of the defect. It is certainly no defence that he has employed a competent independent contractor to check the safety of his premises. The rule is otherwise in regard to natural objects on one's land, such as trees. In that case it is in practice necessary for the plaintiff to prove negligence to succeed (*B.R.S.* v. *Slater*[1964] 1 WLR 498), although the standard of care required of the defendant may be a high one, for example, the employment of skilled advisers (*Quinn* v. *Scott* [1965] WLR 1004). Free passage along a highway must not be obstructed.

(b) *To passers-by on a public or private right of way.* Such a right of way must not be obstructed but otherwise the owner or occupier of land over which it passes generally owes no duty of care to those using it (*Greenhalgh* v. *British Railways Board* (1969) 2 QB 286; *Holden* v. *White* (1982) QB 679 compare *Thomas* v. *British Railways Board* (1976) QB 912). However the law on these topics is complex and the position may vary with the particular facts. Accordingly specialist legal advice should be sought.

(c) *To an employee or lawful visitor.* The employer's liability to his employee is based on negligence, but the standard of care required is high. In certain premises special statutory rules may apply, for example, The Factories Act 1961, and regulations made thereunder.

The liability of an occupier of premises to his lawful visitor is governed by The Occupier's Liability Act 1957. The duty owed is the common duty of care, that is, a duty to take reasonable care having regard to all the circumstances. The duties of owners and landlords are also governed by the Defective Premises Act 1972.

(d) *To operatives who borrow equipment found in premises where they are working.* Basically the owner must take reasonable care to see that the equipment is safe for the purpose to which it is reasonably foreseeable that the operative will put it. The standard of care is to be judged in the light of all the relevant circumstances, including the experience and technical knowledge of the owner and the operative.

If an owner leaves equipment, which he knows or should know is defective, in a place where as a reasonable man he should foresee that the operative may find and use it, and the operative uses it and because of the defect is injured, the owner is liable for his negligence.

(e) *To neighbours.* The owner or occupier of land or buildings is under a duty not to injure his neighbour or harm his property negligently or by a nuisance. In practice in most cases if liability exists it will do so in negligence and nuisance. Nuisance consists essentially of unreasonable use of one's premises which interferes with one's neighbour's reasonable enjoyment of his own. It may arise from such varied sources as noise, smell, subsidence caused by encroachment of tree roots, or landslip (*Leakey* v. *National Trust* [1980] QB 485). The rules for different types of nuisance differ but generally no liability exists unless the defendant created the nuisance or knew or ought to have known of its existence (*Sedleigh-Denfield* v. *O'Callaghan* [1940] AC 880). What a person ought to know, or what is reasonable, are judged by the standards of the day. Thus it will usually be held that a property owner or occupier ought to know of the danger of subsidence caused by tree root action and ignorance of it would be no defence (compare *Solloway* v. *Hants C.C.* (1981) 79 L.G.R. 449). Liability is stricter in the case of withdrawal of support by building or similar operations. There is a right of support for land in its natural state, but not for buildings on the land unless such a right exists as an easement, generally by the passage of time, usually twenty years. If such right exists it is infringed by withdrawal of support causing damage, and ignorance that this would occur, or the fact that the work was done by a reputable independent contractor, are no defence.

(f)   *To trespassers*. A limited duty of care is owed by occupiers to trespassers. It cannot be stated briefly and comprehensively and can only be touched on here. The duty is less than one of reasonable care (owed to lawful visitors) but more than merely avoiding recklessness or intentional injury. It has been called a duty of ordinary humanity: *British Railways Board* v. *Herrington* [1972] AC 877. Its nature and extent depend on the facts of each case and include such matters as that children are less able to protect themselves than adults so allowance must be made for this.

(ii) *Right to Light* (sometimes called 'Ancient Lights') is an easement enjoyed by the owner of a building on one piece of land (the dominant tenement) against the owner of the adjacent land (the servient tenement). It is a right to receive light through any window of the building, acquired by not less than twenty years' enjoyment. The corresponding duty of the owner of the servient tenement is not to refrain from diminishing the quantity of light being enjoyed through the window but to refrain from diminishing it to such an extent as to be a nuisance, that is, an interference with the comfortable use and enjoyment of the building according to the ordinary notions of mankind. A building owner who has acquired an easement of light can require an adjoining owner to fell or lop a tree restricting light to his property, if it constitutes a nuisance for which the adjoining owner is liable. When an owner acquires a property where a tree or trees on adjoining land are obstructing light to the newly acquired property, he can only require his neighbour to fell or lop his trees if an easement of light already exists in respect of the property purchased (twenty years' enjoyment or more) and the interference is a nuisance. If the interference has been there a long time it may be difficult to prove twenty years' enjoyment of light in the amount now sought.

(iii) *Entry on to another's land*

(a)   *To maintain one's own property*. There is no right to do so but the Law Commission has recommended the creation by statute of such a right. Generally speaking, any entry on to the land of another without his prior consent is a trespass.

(b)   *To comply with a Defective Premises or a Dangerous Structures Order*. (i) *Outside London*. It would seem that under the Building Act 1984 a Local Authority may go on to property next door to defective or dangerous premises (but under separate ownership) to remove the defect or danger without the consent of the owner of that adjacent property (s.s. 76 & 78) but that the owner of the premises against whom an order is made has no such power. This is because the statute authorises the Local Authority to "execute such works" or "take such steps as may be necessary" to remove the danger, whereas an order only requires the owner to take such steps, and there is no provision increasing his civil rights. In this latter case, if the owner fails to obey the order, the Local Authority may carry out the work itself and recover the cost from the owner. The innocent neighbour to the defective or dangerous structure is entitled to full compensation for any damage caused by the authority's exercise of its powers: s.106. (ii) *In London.* By similar process of reasoning the position seems to be the same under the London Building Acts (Amendment) Act 1939.

(c) *To abate a nuisance to one's own property caused by action or want of action on the part of one's neighbour.* The right to take the law into one's own hands by abating a nuisance must be exercised to the minimum extent necessary to achieve this object. Thus if the nuisance can be abated without going on to the offending neighbour's land, for example, by lopping protruding branches, one must not go there.

It will often be safer and more effective in the long run only to proceed under the protection of a court order, for the court can and will then authorise all steps necessary for the enforcement of its order, including if necessary committing a recalcitrant landowner to prison for contempt and allowing the victim to go on his land to take appropriate remedial action.

(iv) *General liabilities of others to a property owner*

For practical purposes these have largely been dealt with in parts (i)(e) and (ii) above in this section. The special position of statutory authorities (for highways, electricity, gas, telephone cables, and the like) is too complex to be dealt with here. One aspect does warrant special mention. If a Local Authority's building inspector negligently fails to discover defects in a building being built which involve breach of the Building Regulations, the Local Authority is liable to the occupier (but not a bare owner) of the building when damage to it occurs as a result, but in general no such liability exists where the breaches of the Building Regulations are the fault of the building occupier himself or his professional advisers: *Anns* v. *Merton London Borough Council* (1978) AC 728; *Peabody Fund* v. *Parkinson Ltd* (1985) AC 210; *Investors in Industry Ltd* v. *S. Bedfordshire D.C.* (1986) Q.B.1034.

Limitation of action

Rights of action are barred by passage of time, three years in the case of personal injury or death, six years in the case of property damage or loss. For breach of a contract under seal the period is twelve years, which shows the advantage of making such contracts.

Time runs from the date when the cause of action arises. In tort this is generally when the injury or damage occurs. But in cases of personal injury or death, excusable ignorance of the facts giving rise to the right of action may postpone the running of time:Limitation Act 1980 s.s. 11–14 and 33. In cases of property damage, such as defects in a negligently built building, the law was that the right of action arose when the damage came into existence and not when the property owner discovered or could reasonably be expected to have discovered the defects: *Pirelli* v. *Oscar Faber & Partners*, 1983 2 AC 1. With effect from 18th September 1986 the law has been altered by the Latent Damage Act 1986:in negligence cases involving latent damage the limitation period expires three years from the date when the plaintiff discovered or could reasonably have discovered the damage if this gives a longer period than six years from the occurrence of the damage, but in no case may an action be brought more than fifteen years after the date of the alleged negligence.

In contract, time runs from the date of the breach, which may be long before any damage occurs, for example, an architect who negligently designs inadequate foundations.

But where a duty of care exists in contract a claim in tort may now also be made (*Batty* v. *Metropolitan Property Realisations Ltd* (1978) QB 554; *Midland Bank Trust Co. Ltd* v. *Hett, Stubbs & Kemp* (1979) Ch. 384) so in such cases the distinction is academic.

In cases of negligent survey the damage occurs and time runs from the date when the Client acts on the report, for example, acquires the property: *Secretary of State for the Environment* v. *Essex Goodman & Suggitt* (1986) 1 WLR 1432.

If a person is a minor or of unsound mind when a right of action accrues to him, time does not run against him until he ceases to be under such disability.

The running of time may be postponed by fraud or deliberate concealment. Time then does not run until the plaintiff discovers or reasonably could discover the fraud or concealment. If a builder consciously covers up inadequate foundations by going on building over them he is guilty of deliberate concealment.

Time ceases to run once an action has been started, by issuing a Writ. But even then the action must be proceeded with by the plaintiff at a reasonable pace or the defendant may have it dismissed for want of prosecution.

Damages and other remedies

The main remedies with which a surveyor may be concerned are damages and injunctions.

(i) *Damages*

This means financial compensation. The measure of damages is basically the amount of money necessary to put the injured party in the same position as if the tort had not been committed or the contract had been performed (not "the breach of contract had not occurred").

Thus the measure of damages recoverable by the client for a negligent survey will usually be the difference in value not the cost of repair since if the survey had been properly carried out and the defects discovered this would usually result in the vendor reducing the price by the difference in market value of the house with and without the defects (not reducing it by the amount of the cost of their repair): *Philips* v. *Ward* (1956) 1 WLR 471; *Perry* v. *Sidney Phillips & Son* (1982) 1 WLR 1291. But as a general rule, courts in awarding damages will seek to apply the fundamental principle of proper financial compensation referred to at the outset, and will not be side-tracked from it by rules that may appear to have grown up around it over the years: *Dodd Properties Ltd* v. *Canterbury City Council* (1980) 1 WLR 433.

Substantial damages may also be recovered for inconvenience and mental distress: *Bunclark* v. *Hertfordshire County Council* (1977) 243 E.G. 455; *Perry* v. *Sidney Phillips & Son, supra*.

Repair of damaged property may result in improvement through the inevitable replacement of old by new. Damages are not to be reduced on that account but only if the plaintiff, whilst executing reasonably necessary repairs, takes the opportunity of effecting additional improvements—their cost is not recoverable: *Harbutt's Plasticine* v. *Wayne Tank & Pump Co.* (1970) 1 QB 447.

Delay in effecting repairs will in inflationary times greatly increase their cost but if the plaintiff has acted reasonably in deferring them he will recover the increased cost, for example, where his resources are limited and he wants to wait and see how much if anything he will recover from the defendant before deciding what repair work to do: *Dodd Properties Ltd* v. *Canterbury City Council, supra.*

Generally speaking, only the person who has suffered the relevant damage can recover damages for it. Thus if someone buys a house with settlement damage due to tree root action, such damage should be reflected in a reduced price and only the owner at the time when the damage occurred can sue for it. In some cases this may mean that the person who suffers the loss cannot sue for it and the person who can sue for it suffers no loss. A court may well then allow the former a right of action: *Masters* v. *Brent London Borough Council* (1978) QB 841. An alternative solution will be for the person actually damaged to take an assignment of the right of action of the undamaged person possessing it: *Trendtex Trading Corp.* v. *Credit Suisse* (1982) AC 679.

Interest on damages will normally be awarded, generally from the date of making the claim or serving the writ to judgment. The rate may be bank overdraft lending rates over the relevant period, or something less than this. A judgment itself carries interest until the date of its satisfaction.

(ii) *Injunction*

This may be granted to compel a defendant to remedy his wrong (for example, to pull down a wall built on his neighbour's land) or to stop committing a wrong (for example, a noise nuisance, or unlawful building operations). Even if no wrong has yet been committed, if the plaintiff can show that he will sustain substantial damage from what the defendant is about to do unless restrained by court order, then the court will grant a *quia timet* injunction.

Appendix A: Wood-boring Insects

Timber does not contain the seeds of its own destruction for it will last indefinitely unless attacked by an outside agent such as wood-rotting and sap-stain fungi (discussed in appendix B). Most other damage to timber is caused by insects and of these, a few species of wood-boring beetles are by far the most serious pests in temperate regions. In the sub-tropics and tropics, termites, popularly called white ants, can be devastatingly destructive, and in salt and brackish water several species of crustaceans can be extremely destructive. Not all insects associated with timber cause damage, for example, the land form of wood-louse, found under any piece of wood that has been left in contact with the ground, in sheds, or in the open for any length of time, causes no significant damage.

In temperate climates the timber user is concerned with a small number of beetles which, like all insects, pass through several stages of development from egg to adult beetle, called a life-cycle. The adult beetles mate, and the female lays eggs. These hatch into grubs, become larvae, which go through a resting stage, followed by pupation, when metamorphosis occurs, to eventually emerge as an adult beetle. The damage beetles do is frequently referred to as 'worm' in timber. Except for the group known as *Ambrosia* beetles, the damage is done by the feeding larvae, which tunnel in timber, feeding on cell-wall substance or the cell content. With *Ambrosia* beetles it is the adult that does the tunnelling, introducing a fungus, known as *Ambrosia fungus*, on which the larvae feed.

The different types of insects, their means of identification, and methods of control, are discussed fully in various pamphlets and bulletins. *BRE Digest 307*: 'Identifying damage by wood-boring insects' and the publication *'Timber pests and their control'* produced by TRADA and the British Wood Preserving Association, are excellent publications on the subject of beetle infestation.

Beetle identification

Basically, the size and shape of the exit holes, and the nature and colour of the bore dust (frass) provide the means of identifying the different pests. Unfortunately for identification purposes, it is by no means certain that adult beetles, or the remains of such beetles, will be found even when there is continuing active attack. Some beetles attack timber in the forest or while it is seasoning after conversion, whereas others only attack dry timber, or timber that has first been decayed by wood-rotting fungi.

AMBROSIA BEETLES. This group does a considerable amount of damage to tropical timbers in particular, although there are a few species in this country that are occasionally found in oak and other hardwoods. The popular names for these beetles is 'pin-hole' or 'shot-hole' borers. The adult beetle tunnels spirally at right-angles to the grain, into living trees and newly felled logs, laying the eggs in specially constructed egg-chambers. At the time of egg-laying, the adult introduces the *Ambrosia fungus* into the galleries, upon which the larvae feed when hatched. Provided the extent of pin-hole and shot-hole borer damage infestation is not excessive, the timber can safely be used as there is no risk of spread of attack and still less risk of infections of other timber in the vicinity. This is because as the converted timber dries, the fungus also dies, and the larvae cannot complete their life-cycle for want of food. It is important to recognise such wormy timber for what it is, since there is no need to apply an insecticidal treatment to timber containing this form of beetle attack, see figure 23. Among carcassing timbers, hemlock is the one species likely to contain pin-worm damage, and since hemlock is quite extensively used in modern estate development, the surveyor should be able to recognise the holes for what they are, so that he can assure his client that the attack is of no practical significance, and calls for no remedial measures.

LONGHORN BEETLES. A large number of species belong to this group, the majority of which are tropical whose larvae usually die when the timber is brought into temperate climates. Exceptionally, an adult beetle may emerge a few years later, but it would be unable to mate and re-infest timber in a temperate climate. There is one temperate-region species, the house longhorn beetle *(Hylotrupes bajulus L.)* figure 24(i). This is a serious pest in parts of northern Europe and in some districts infestation has to be notified, and appropriate remedial measures are compulsory. The house longhorn beetle which is 10–20 mm (approximately $1/2$–$3/4$ inch) long, attacks only the sapwood of softwoods. The larvae tunnel just beneath the surface, completely destroying the sapwood behind a shell of apparently sound wood. The life-cycle is a long one and, in some cases, can be upwards of ten years. The result of such a long life-cycle is that serious damage may result before the first flight holes bring the attack to light. The frass contains pellets that resemble the flints of petrol lighters when examined with a pocket lens, see figure 24(ii). The flight holes made by the adult beetles as they emerge are 6–10 mm (approximately $1/4$ to $1/2$ inch) long and oval in section. The finding of frass or oval flight holes is not proof of house longhorn infestation, and still less that there is continuing activity. Other beetles produce similar frass and oval flight holes. Active house longhorn beetle infestation is generally found in the Home Counties. Even in these areas no active infestation has been found in any building more than fifty years of age, unless new timber has recently been used for repairs in such buildings. The reason for this appears to be that the habitat requirements of the house longhorn beetle are critical. The larvae only hatch if temperatures are sufficiently high and because of the long life-cycle, it is probable that re-infestation often cannot occur because

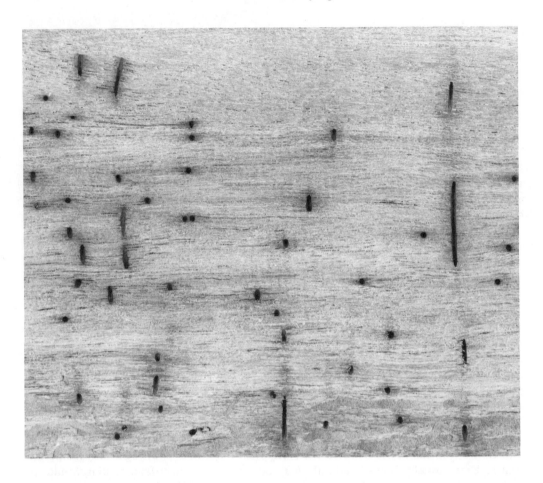

Figure 23   Pin-hole damage (*Crown copyright: Reproduced by permission of the BRE Princes Risborough Laboratory*)

a sufficient number of male and female beetles have not emerged at about the same time, and during a spell when temperatures are adequate for the eggs to hatch. Although attack is not confined to roof timbers, it is most frequently found in rafters, purlins and ceiling joists within the roof void, the particularly vulnerable timbers being rafters on either side of a chimney stack containing flues in regular use. Exit holes in the lead flashings to dormer windows have been found indicating that the adult's jaws are powerful.

If the surface of carcassing timber is seen to be bulging outwards as if the wood were covered with blisters, attack should be suspected and the timber should be probed, when a shower of typical frass is likely to spill out. This may be the only indication that attack is in progress if a life-cycle has not been completed and there are no characteristic oval exit holes. When attack is found the surveyor must ascertain whether or not the infestation is still active. If the attack is active, drastic remedial measures are called for. The attacked sapwood should be cut back to sound wood and the whole roof void should be given an insecticidal

(i)   (ii)

Figure 24   (i) House longhorn beetle and (ii) frass (*Crown copyright: Reproduced by permission of the BRE Princes Risborough Laboratory*)

treatment. If the proportion of sapwood is high, the damage may sometimes be so serious as to call for complete replacement of roof timbers. This was found to be necessary in a large number of council properties in the Camberley area. In several districts in Berkshire, Hampshire and Surrey, the 1976 Building Regulations called for the special pretreatment of all roof timbers in new buildings; this is not the case in the 1985 Regulations which give no specific advice on the measures to be taken to protect the timber.

In endemic areas pressure-treated timber, or timber treated with boron by the diffusion process, is much to be preferred to a dip or brush application with organic solvent preservatives. *BRE Information Paper IP 12/82* 'House Longhorn Beetle Survey' provides information of the damage caused by the beetle and gives advice on preservative treatments.

POWDER-POST BEETLES. There are two families of these, the *Bostrychidae* and the *Lyctidae.* The former are mainly tropical species whereas the latter are both tropical and temperate- region species. The larvae of these two families do not live on cell-wall substance, but on the starch content of the sapwood of certain timbers. Unlike the sap-stain fungi however, the larvae have to destroy the cell-wall substance to get at the stored starch. The egg-laying habitats of the two families of powder- post beetles differ, as do their demands in regard to the degree of dryness of wood favoured for egg-laying. The *Bostrychid* beetles bore into wood, constructing a Y-shaped egg tunnel, which is kept free from dust and in which the female lays her eggs. When the eggs hatch, the larvae tunnel into the wood longitudinally, packing the gallery system with fine, flour-like dust. *Bostryhid* beetles will attack timber soon after conversion, while boards, planks, and scantlings are still in stick. In practice the only appropriate remedial measure is to remove the sapwood. The galleries of the Bostrychid beetles are up to 5 mm (approximately ³/₁₆ inch) in diameter. Attack occurs in the sapwood of imported hardwoods, but dies out within a year and hence does not call for any insecticidal treatment.

(i)   (ii)

(iii)

Figure 25   (i) *Lyctus* beetles, (ii) damage and (iii) frass (*Crown copyright: Reproduced by permission of the BRE Princes Risborough Laboratory*)

(i)   (ii)

Figure 26   (i) Common furniture beetle and (ii) frass (*Crown copyright: Reproduced by permission of the BRE Princes Risborough Laboratory*)

Lyctus beetles, figure 25(i), are typically pests of seasoning yards, but re-infestation may occur if timber containing active infestation is utilised for such purposes as furniture, panelling and flooring. Except for a reported case of attack in a South African pine, only certain hardwoods are subject to *Lyctus* infestation. Attack is usually confined to the sapwood, although the adult may emerge through heartwood adjacent to the sapwood. Typical *Lyctus* damage is shown in figure 25(ii). The size of the vessels or pores in the wood is a limiting factor in the beetles' reproduction, for they must be large enough to admit the ovipositor (egg-laying tube) of the adult female, as it is in the vessels that the eggs are almost invariably laid. Such fine-textured timbers as beech are immune. Several timbers with vessels large enough to permit egg-laying are immune because the sapwood does not contain sufficient starch for the beetle to select the species for egg-laying.

The life-cycle of *Lyctus* from egg to adult beetle is normally about one year, but the period may be as short as ten months and, where food supplies are deficient, the life-cycle may be considerably extended to two, or even three to four years. Adults normally emerge from April to September, appearing in largest numbers in June, July and August. Immediately on emerging the adult beetles, which are red-brown in colour, and 2–6 mm (approximately $1/16$ to $1/4$ inch) in length, proceed to mate and the female begins egg-laying. She is most fastidious in regard to the suitability of the particular piece of wood selected for egg-laying, for it must be rich in starch. The flight hole is circular in outline and about 2 mm (approximately $1/16$ inch) in diameter. The frass has the consistency of flour when rubbed between the fingers, figure 25(iii), and it has no definite structure.

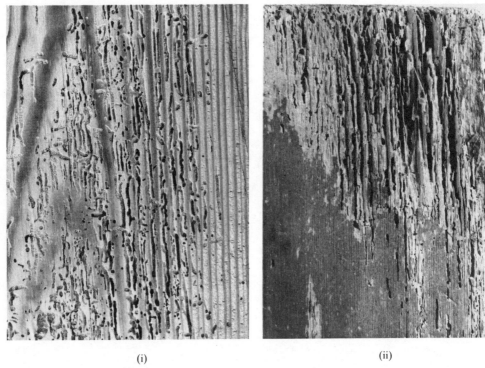

(i)   (ii)

Figure 27 (i) Common furniture beetle damage and (ii) weevil damage (*Crown copyright: Reproduced by permission of the BRE Princes Risborough Laboratory*)

There is another beetle very similar to *Lyctus* in size, shape and colour, namely the common grain beetle. These two beetles can readily be distinguished by the number of clubs at the end of the antennae, in *Lyctus* there are two clubs, whereas the grain beetle has three.

Control of *Lyctus* infestation should be done in the seasoning yard, as it is extremely difficult to carry out an adequate *in situ* chemical treatment of attacked timber. This is because very early in the attack, the larvae tunnel in all directions reducing the wood to a fine powder, which blocks the path of wood preservatives applied *in situ*. The sapwood of carcassing timber, in period houses in particular, is likely to have been completely destroyed in part by *Lyctus* attack, but subsequently by the common furniture beetle. Such attacked timber calls for no remedial measures because the attack will have been dead for many years, and all the surveyor has to do is to satisfy himself that there is sufficient timber left for structural purposes. In modern developments, *Lyctus* infestation is most likely to be found in the sapwood of oak flooring and joinery, and here the only satisfactory solution is to recommend the renewal of all timber containing sapwood.

COMMON FURNITURE BEETLE, Anobium punctatum. This has received much more attention in recent years than its importance warrants, and tens of thousands of pounds have been spent on general insecticidal treatments quite unnecessarily. The damage is done by the larvae which hatch from eggs laid in cracks in the wood, in joints of made-up woodwork and, more rarely, in old flight holes. The larvae travel along the grain, but as they feed and grow, they tunnel in all directions, filling their galleries with loosely packed granular frass which feels gritty when rubbed between the fingers. The individual pellets making up the frass are appreciably thinner than those in longhorn beetle frass, see figure 26(ii). The adult beetles are between 2 and 6 mm (approximately $^1/_{16}$ to $^1/_4$ inch) in length and leave a 2 mm ($^1/_{16}$ inch) exit hole. The head is enclosed in a characteristic hood-shaped thorax and the wing cases are pitted, hence the species name *punctatum* see figure 26(i).

The life-cycle from egg to adult beetle may be as little as one year, but it is now thought that, on average, it is about three years. Adult beetles emerge in May, June, July and August, and mate. The females lay their eggs in suitable places but they will not lay on smooth surfaces.

The common furniture beetle is widely known as a pest of old furniture and of hardwood constructional timbers in period houses and, more recently, it has been recognised as a common pest in the sapwood of softwoods in buildings of all ages. It was formerly thought that initial attack did not occur until the timber had been in service for several years, and that it was necessary for the timber to have 'matured' in some way for it to become attractive to the beetle. Entomologists now conclude that initial infestation may occur as soon as the timber has become seasoned, but the presence of attack may not be discovered until several life-cycles have been completed, and flight holes are quite numerous.

In small articles of furniture the damage is not confined to sapwood, whereas in beams and constructional timbers, it usually is. In furniture and small wooden articles the damage done may be quite serious, as for example, attack in the leg of a chair. In structural timbers, attack is mainly confined to the sapwood, and structural damage is only serious when the amount of sapwood is abnormally high. Attack in softwood timbers may develop to the extent of causing some timbers to collapse if sufficient sapwood is present, but the presence of a beetle population is probably of more importance because of the risk of subsequent infestation of furniture. It is important for the surveyor to be able to recognise this type of beetle infestation and also to determine whether the infestation is active, before recommending a chemical treatment. Typical furniture beetle attack is shown in figure 27(i). It is recommended to refer to it as the common furniture beetle, or woodworm, as the scientific name may well not be known to the client.

DEATH WATCH BEETLE, Xestobium refuvillosum, lays eggs in crevices, cracks, or old exit holes, and the larvae create the damage by tunnelling and feeding on the wood. Attack is usually confined to old timbers of several species of hardwoods, but it has been known to spread to adjacent softwood timbers or even to be found in a house containing no hardwoods. Attack is not confined to the sapwood, but is more likely to begin in sapwood than in heartwood. Adequate moisture and the presence of fungal decay are conditions favourable for infestation. The galleries made by the larvae are about 3 mm (approximately $^1/_8$ inch) in diameter and they are filled with coarse frass, containing bun- shaped pellets, figure 28(ii). Removal of decayed wood and the causes of decay are the first essential steps in eradicating this beetle. Frequently, however, major structural damage will have already occurred, when replacement of the attacked timbers rather than *in-situ* chemical treatments, is the only practicable course. *BRE Information Paper IP 19/86* gives further advice on controlling death watch beetle. Typical death watch beetle damage is illustrated in figure 28(i). The adult beetle is brown or red-brown in colour, and varies from 6 to 8 mm (approximately $^1/_4$ to $^5/_{16}$ inch) in length. The thorax is smooth, broad and flanged, and there are patches of short yellow hairs on the wing cases which give the beetle a mottled appearance. Moreover, if the beetle has only recently emerged

(i)

(ii)

Figure 28   (i) Deathwatch beetle/damage and (ii) frass (*Crown copyright: Reproduced by permission of the BRE Princes Risborough Laboratory*)

(i)   (ii)

Figure 29   (i) Wood-boring weevils and (ii) frass (*Crown copyright: Reproduced by permission of the BRE Princes Risborough Laboratory*)

there is likely to be frass adhering to the wing cases and head. Under optimum conditions the life-cycle may be only one year, but in less favourable conditions it is prolonged over two to several years.

Somewhat similar in size and shape to the death watch beetle is *Attagenus pellio Herbst.*, the common carpet beetle. This beetle has the same kind of broad, flanged thorax, with usually two tufts of light-coloured hairs on the wing cases. It is flatter than the death watch beetle, almost black rather than brown, and the thorax is particularly shiny. It moves much more rapidly than the sluggish death watch beetle, this rapid movement giving a clue to its identity.

WOOD-BORING WEEVILS. These belong to the genus *Pentarthrum*. Both beetles and larvae bore in timber, causing damage resembling that of the common furniture beetle, but the galleries are rather smaller in diameter and the frass is rather finer in texture than that of the furniture beetle, see figure 29(ii). It is possible to confirm weevil infestation by finding parts of, or whole, beetles with their characteristic snouts, see figure 29(i). Both hardwoods and softwoods are liable to infestation and attack is not necessarily confined to the sapwood. In most cases a prerequisite of attack is the presence of decay, and hence weevil infestation is most likely to be found in basement and poorly ventilated ground-floor floors, where there is no oversite concrete and fungal decay has become established. The measures called for to dispose of the decay hazard will eliminate the weevil without insecticidal treatment. Typical weevil damage is shown in figure 27(ii).

Bats

The surveyor when advising on a treatment for the eradication of wood boring insects and for the *in-situ* preservation of timbers in the loft must ensure that no protected species are using the roof void as a roosting site. Probably the biggest threat is to bats who are now protected under the Wildlife and Countryside Act 1981 because of their special needs for roosting. If the surveyor believes that bats are present then he can seek advice from the Nature Conservancy Council at Northminster House, Peterborough, PE1 1VA. This organisation also produces a useful guide on the subject called *'Bats In Roofs: A Guide for Surveyors'*.

Appendix B: Decay and Fungi

Whereas trees, shrubs, herbaceous plants, grasses and the like manufacture their own food materials from simple substances such as water from the soil, carbon dioxide from the atmosphere and energy trapped from sunlight, there are a large number of plants called fungi, that feed only on organic matter of either plant or animal origin. Examples of these are the edible mushrooms, the many decorative but often highly poisonous toadstools to be found in our fields and woodland, and moulds that grow on almost any damp, organic material such as bread, fabrics, or wood. Most forms of decay and sap-stain in timber are caused by fungi that feed either on the cell tissue, that is, 'wood substance', or the cell content of woody plants. Some forms of decay are however, chemical or bacteriological in origin, and certain infiltrates present in wood cause staining when such timbers come in contact with metals, for example, iron and steel. It is important to distinguish between the wood-rotting fungi responsible for decay of timber, and those that feed on the cell content causing sap-stain. The former consume certain constituents of the cell walls and lead to the disintegration of woody tissue, whereas the latter feed only on stored plant food material occurring in certain cells of the sapwood, leaving the cellular structure intact. Wood-rotting fungi seriously weaken timber, ultimately resulting in its failure, whereas sap-stain fungi spoil the appearance of wood, but do not affect most strength properties. In effect sap-stain is not a preliminary stage of decay, but it is a warning sign. Scantlings or boards containing an appreciable amount of sap-stain also contain a high proportion of sapwood, which, given the right conditions, may later become attacked either by wood-rotting fungi or certain insects.

All fungi produce fruit bodies or fructifications, the vegetative part of the plant being out of sight in the feeding medium. The fruit bodies of the different wood-rotting fungi are either flat, fleshy or woody plates, the undersides of which bear spores or seeds. The destructive part of the fungus is the vegetative system, or *mycelium*, made up of numerous exceedingly fine thread-like tubes called *hyphae*. These may become matted together to form a felt-like mass. *Hyphae* grow by elongating at their tips, passing from cell to cell of the host plant, feeding on the wall or cell contents in their path. The complete life-cycle of the fungus is therefore, spore—hyphae—mycelium—fruiting body—spore–

A large number of different species of fungi attack wood. Some species attack the living tree if the latter is unhealthy, others, timber in log form, and yet others attack timber that has been taken into service. The fungi of interest to the user of timber are those that attack timber in service. There are several different species of these, but the surveyor is usually only concerned in the UK with under half a dozen species, all of which require five conditions to be satisfied for attack to occur. These conditions are:

(a) A source of infection, which may be either spores or *mycelium*.
(b) A food supply, which is any piece of timber, although some timbers are more resistant to attack than others and the sapwood of all species is much more susceptible than the heartwood.
(c) Suitable temperatures, since all fungi cease active growth below 0°C and growth is very slow between 0°C. and 4°C, accelerating as the temperature rises until the optimum range is reached. The range varies with different species but for the majority of species growing in a temperate climate the optimum range is between 18°C and 24°C. High temperatures are lethal to all fungi, but some species are more susceptible than others to temperatures only slightly over 32°C, if maintained for any length of time.
(d) Oxygen, which is available from the atmosphere.
(e) Adequate moisture, although the moisture requirements of different species vary, and hence the common differentiation into the so-called dry rots and wet rots. The important fungi that attack wood in service grow most vigorously when the moisture content of wood is in the range of 24–28 per cent (moisture calculated as a percentage of the dry weight of wood), but the dry rots can continue active growth so long as the moisture content of the wood is at or above 20 per cent. The wet rots, however, do not survive at these low moisture contents; further, the moisture content of wood can rise to the point where it is too wet to permit continuing attack by the dry rots.

Wood-rotting fungi

Actual decay in timber may be detected by the abnormal colour of the wood, by the transverse fracture of the fibres on longitudinal sawn faces and by lifting the fibres with the point of a pen-knife, when, if the timber is decayed, the fibres will snap instead of pulling out in long splinters.

The two main constituents of wood are cellulose and lignin. Certain fungi feed mainly on the cellulose and are called brown rots, whereas others feed both on cellulose and lignin, and are know as white rots. The different wood-rotting fungi can be further subdivided according to the form the decay takes. In some, the wood shrinks and longitudinal fissures appear, while in others, transverse cracks develop and the wood ultimately breaks into cubical fragments, and yet in others, decay occurs in pockets.

The consumer is not normally concerned with fungi that attack the living tree or freshly felled logs, but if such infected timber is stored close-piled after conversion and taken into service inadequately seasoned, there is a risk that decay will continue to develop. This problem is particularly associated with such timbers as Douglas fir which is normally shipped very soon after conversion and requires to be properly stacked and allowed to season before being used.

Figure 30   Dry rot fruiting body (*Crown copyright: Reproduced by permission of the BRE Princes Risborough Laboratory*)

It is essential for surveyors to be able to recognise the few wood-rotting fungi that commonly attack timber in buildings, because, depending on the species, the appropriate remedial measures vary. By far the most important of these fungi is the true dry rot *Serpula lacrymans* which is a brown cubical rot. The appearance, both of the fungus and of the infected wood depends on the stage attack has reached, and on the growth conditions for the fungus. In damp conditions, particularly in still air, it develops as a white, fluffy, cotton-wool-like mass, spreading over the surface of the attacked wood. In drier conditions the *mycelium* forms a grey-white felt over the wood, usually with small patches of bright yellow or lilac. Branching strands may develop from the felt, varying in thickness from coarse threads to strands as thick as a lead pencil. These strands are made up of hyphae that conduct plant nutrient (food materials) and water. They can penetrate loose mortar in a brick wall and cross steelwork and concrete to reach new feeding grounds to attack dry timber. The fruit bodies are soft, fleshy plates with white margins, figure 30. Numerous folds or shallow pores occur on the surface of the fruit bodies which contain the rust-red spores. These are microscopic and so light that they are easily blown about. They are sometimes produced in such quantities that a whole room may be covered with a layer of rust-red spores. The fruit bodies sometimes grow vertically in the form of a thick bracket, when the pore-bearing surfaces become elongated, like small stalactites. Water may be exuded in drops by the fruit bodies and fluffy growths of *mycelium* in still air, hence the name *lacrymans* or 'weeping' for the commoner species of *Serpula*.

The fruit bodies that grow out into the air and light are frequently the first indication of dry rot in a building, but unventilated rooms or shut-up houses that are infected usually have a characteristic musty odour. Slight wavyness on the surface of panelling, skirtings, linings and other joinery items are further indications of possible extensive damage. Wood beneath

Figure 31   Dry rot damage (*Courtesy of Cuprinol Limited*)

a coating of *mycelium* is wet and slimy to the touch, but in the final stages of attack it is dry and friable, brown in colour, and breaking up into cube-shaped pieces see figure 31.

It is often said that *Serpula* can remain dormant for several years, renewing active growth when adequate supplies of moisture become available, but this is incorrect. If the source of all moisture is cut off so that the medium on which the fungus is feeding dries out to a moisture content of about 16 per cent, *Serpula* will not survive for even a year. In addition, the spores will not be capable of germination if they are exposed to dry conditions for only two to three months. Unfortunately, once attack has become established in a building and the *mycelium* has penetrated deeply into a saturated wall, the latter may well take two or three or more years to dry out. In this period the fungus will survive more or less in a dormant condition, provided there is a small amount of food material available, for example, fixing plugs for pipe nails on the outside of a wall or small fixing blocks for skirtings and the like on the inside face. When further moisture becomes available, growth is resumed and the *mycelium* continues its travels in search of new feeding grounds.

Coniophora puteana, commonly called the cellar fungus is also a brown rot, but in the final stages of attack the wood sometimes develops longitudinal splits or cracks, see figure 32, and the decayed timber may break up into cubical pieces. This is virtually indistinguishable from the final stages of *Serpula* attack, except that the wood is not permeated with *mycelium*. The decayed timber is dark in colour and extremely brittle. It is capable of being reduced to powder if rubbed between the fingers. Although *Coniophora* grows most vigorously in the moisture range of 24–28 per cent, initial infection usually occurs in conditions of somewhat higher moisture content, too high for the so-called dry rot. *Coniophora* is particularly liable to occur wherever there is persistent leakage of water or condensation. The *hyphae* are always fine and they rapidly turn brown or almost black. Spreading over the face of the wall the *mycelium* resembles the maiden-hair fern, but without leaves. The fruiting bodies are thin, olive-

Figure 32   Cellar fungus damage (*Crown copyright: Reproduced by permission of the BRE Princes Risborough Laboratory*)

green plates, but these are rarely encountered, even where attack is vigorous and of long standing.

Paxillus panuoides is a brown rot requiring very moist conditions. The hyphae are paler than those of *Caniophora puteana*, and the *mycelium* is rather fibrous and yellow or violet. The fruiting bodies, which are often bell-shaped, are olive-green with deep gills on the under surface. The fungus is most likely to be found in a roof void under a leaking gutter.

Serpula and *Coniophora* are by far the commonest fungi likely to be encountered in buildings, and the first named is unquestionably the most serious of all wood-rotting fungi, being responsible for untold damage annually. If an attack has ceased and there are no remains of fruiting bodies, it is sometimes difficult to identify the fungus responsible for the decay. With dry rots, the decayed timber will be found to be permeated with strands of *hyphae*, usually visible to the naked eye, but certainly distinct with a pocket lens.

Mention should also be made of two fungi, unimportant in themselves, that are nevertheless an indication that dangerously damp conditions exist. Elf cups (*Peziza* sp.), which are yellow-brown cups about 25 mm (1 inch) in diameter, sometimes develop on plaster ceilings following flooding from defective plumbing or frost damage. Unless steps are taken to secure rapid drying out of the affected areas there is a risk of subsequent wet or dry rot infection. Another fungus that should be similarly regarded is a species of inky cap (*Coprinus* sp.). This fungus, which produces small soft toadstools that dissolve into an inky fluid, is often found growing on damp cellar walls, but it may also appear on the underside of ceilings saturated by persistent plumbing leaks or defective internal gutters.

Sap-stain or blue-stain in timber is caused by several species of fungi of the mould type. These fungi are distinct from those that caused decay, hence blue-stain is not an incipient state of decay, but its presence may be an indication of conditions favourable for the attack of wood-rotting fungi. Moreover, badly blued timber should be suspected of containing incipient decay or dote as a result or a separate infection by a wood-rotting fungus. All staining of wood is not necessarily blue-stain. Green timber rich in tannins that comes in contact with iron, when worked, may become stained blue-black. This is the result of chemical action and the stained areas are usually superficial and easily planed off. Similarly, coloured stains caused by sap-stain fungi quickly penetrate deeply into wood. Besides the fungi responsible for 'blue-stain', there are several other mould fungi that stain wood green, pink, purple and, more rarely, brown. The majority of these produce a powdery or downy growth of mould that is easily brushed or planed off.

The five conditions essential for the development of wood-rotting fungi are also essential for sap-stain fungal infection, except that higher initial moisture contents are necessary and the food of sap-stain fungi is not wood substance itself, but stored plant food material. The right type of food material in sufficient quantities is the limiting factor. In most softwoods these requirements are only found in the sapwood, but in Sitka Spruce, attack may spread to the heartwood. Retention of the bark, saturated with moisture, also inhibits mould growths because of lack of air. Relatively high temperatures are necessary for active growth, the optimum being 21°C to 27°C, and below the optimum, growth is very slow. In temperate regions these temperature conditions only exist in the summer months. Reduction in moisture content in the surface layers of wood can rapidly become a limiting factor as the fungi require moisture contents above the fibre saturation point of the wood, equivalent to a moisture content of about 30 per cent, to initiate the attack.

Once timber has been dried below the critical stage for the blue-stain infections, these fungi should not constitute any problem to the consumer. If conditions in service become such that fresh blue-stain infection could occur, conditions favouring the attack of wood-rotting fungi would also exist, presenting a much more serious problem, and one requiring drastic and immediate remedial measures.

Appendix C: Specimen Reconnaissance Survey Report

This report is made following a survey carried out on Monday 25 March 1986, in accordance with instructions received from Mr F. D. Best, of 'Thistle Doo', Milton Road, Little Bufford, who is proposing to purchase the property. We are asked to inspect the property in as much detail as possible and to report as to the condition of the structure and services. Some floor coverings are laid, and it is agreed that we are not required to lift fitted floor coverings, to pull up or take down any part of the structure for detailed investigation nor to test any of the services. We have not inspected woodwork or other parts of the structure which are covered, unexposed or inaccessible and we are therefore unable to report that such parts of the property are free from fungal rot, beetle or other defects. Our report is presented on this basis and is intended solely for the information of our client and his professional advisers in the present transaction.

(For obvious reasons the addresses in this specimen report are fictional. It is important to make sure that the addresses both of the property to be surveyed and of the client are correct and accurate, including the post codes, which can quite easily be discovered in the majority of cases. If the surveyor cannot get the address right, what chance has he of getting anything else right?)

We understand that our client has already obtained estimates for the treatment of damp and woodworm in the building, as required by the Building Society to whom an application for mortgage has been made. He has also obtained estimates for the installation of a modern central heating system based on a solid fuel boiler.

GENERAL DESCRIPTION

1.0   The property comprises a semi-detached private dwelling house. It has been formed by the conversion of three cottages in a block, and we suggest that this was originally two cottages, and that the third one (at the North East end) was an addition at a later time. The property is semi-detached to the extent that it butts up against the neighbouring property 'Long Thatch', although apart from this the two properties are not connected in any way. We think that the original pair of cottages was probably built around 1870, and that the third one was added not very long afterwards. The conversion to a single dwelling house has taken place fairly recently. We believe that the present vendor is a 'do-it-yourself' enthusiast, and that much of the work was done with his own hands.

SITUATION

2.0   The property is situated on the Northern edge of the village of Little Bufford. As our client lives close by and has done so for many years and knows the district well, we therefore think it unnecessary to include further descriptive details as to the situation of the property, in the course of this report.

2.1   The road to which the property has a frontage runs approximately South West to North East, so that the house is on the North West side of the road, the front elevation facing approximately South East. This convention will be followed in our report and where in this report room measurements are given, the dimension first given in each case is in the North East to South West direction. These dimensions are approximate only and are given in metric and imperial units.

TENURE

3.0   We are advised that the tenure is FREEHOLD and we are not aware of any rights of way, easements or other restrictions which would affect the enjoyment of the freehold title. The solicitor should be requested to check very carefully on this point.

ACCOMMODATION

4.0   The accommodation comprises

4.1   Ground floor

Entrance Porch (on the North East flank)
Entrance Hall   with meter cupboard
Lounge   4.06 x 3.68 metres (13″4′ x 12″1′)
Dining Room   3.81 x 3.71 metres (12″6′ x 12″2′) with modern tiled fireplace and small cupboard alongside
Kitchen   3.73 x 2.64 metres (12″3′ x 8′8′) with inset stainless steel sink unit and fitted cabinets

4.2 First floor

Landing	with heated linen cupboard
Bedroom 1(front)	3.81 x 3.71 metres (12″6′ x 12″2′) with built-in cupboard
Bedroom 2(rear)	adjoining Bedroom 1, 3.73 x 2.46 metres (12″3′ x 8″1′)
Bedroom 3	3.73 x 3.63 metres (12″3′ x 11″11′)
Bedroom 4	4.11 x 3.66 metres (13″6′ x 12″0′)
Bathroom WC	with pressed steel panelled bath (H & C), pressed steel wash hand basin (H & C) and low level WC

4.3 Outside

| Garage | built into the main structure, 3.81 x 6.10 metres (12″6′ x 20″) (this is the ground floor accommodation of one of the original cottages) |
| Garden Shed | of timber construction, 2.40 x 1.80 metres (8″ x 6″) with a pitched felted roof |

DESCRIPTION

5.0 The house is of conventional brick construction, in a style commonly associated with country artisan type buildings. Main walls are 215 mm (one brick) thick and the bricks are laid to Flemish bond. Many alterations have been carried out to the outside walls and a variety of bricks have been used, but the original bricks are probably red hand-made bricks, manufactured locally. The whole of the elevations are painted with a pale green paint, which we think is 'Sandtex' or something similar. We could not identify a damp-proof course in the main walls, although the vendor has stated that he has inserted a chemical injection damp-proof course and there is some evidence to suggest that this had been done. However we could not see clearly the holes where the injection has been made, nor does there appear to be any documentary evidence relating to the work.

5.1 Internal walls are of brick on the ground floor. Most of them are plastered and decorated with wallpaper. The chimney breast in the dining room is clad with vertical tongued and grooved boards and decorative plywood lining is fixed to some of the walls in the kitchen. On the first floor, the walls which run from front to rear are of brick, and the wall between Bedrooms 3 and 4, which on the ground floor is the wall between the Lounge and Dining Room and between the Kitchen and Entrance Wall, is 215 mm (one brick) thick. This is one of the features which leads us to think that this is an original external wall and that the cottage at the North East is a later addition. The other walls on the first floor, which run parallel to the road, are of timber studs with a lath and plaster finish. Most wall surfaces on the first floor are decorated with wallpaper, but part of the wall between Bedrooms 1 and 2 has the plaster stripped away so that the details of the construction can be seen quite easily.

5.2 The main roof has its ridge parallel to the road. There is a gable at the South West end and a hip at the North East end. This roof is of timber construction. Almost certainly it was originally slated but in the fairly recent past it has been stripped and covered with interlocking concrete tiles. These are Marley 'Ludlow' tiles (or similar) in a dark brown colour. Ridge and hip tiles are half round, and there are mild steel hip irons at the lower ends of the hips.

5.3 There is a single large chimney stack passing through the ridge of the roof and serving the fireplace in the Dining Room. It is in red bricks, similar to those which were probably used originally on the lower walls, and the opening in the roof covering where it passes through is protected by a fillet of cement mortar. At the South West end of the house there is a chimney which serves 'Long Thatch', passing up on the gable of the property now under report. This also has a mortar fillet to protect the opening in the roof covering.

5.4 The ground floor is solid, and we think that it has probably been completely renewed. In the Entrance Hall and Lounge it is finished with wood mosaic blocks. Carpet is laid in the Dining Room on a fine cement and sand screed and vinyl sheeting is laid in the Kitchen.

5.5 The first floor is of normal timber construction. It is finished with softwood boards and some carpet is laid.

5.6 The original ceilings were of lath and plaster, and these appear to remain on the ground floor. In the Lounge and Kitchen the ceilings are decorated with a pebble pattern paper, and in the Dining Room and Entrance Hall, they are decorated with expanded polystyrene tiles. Most of the ceilings on the first floor appear to have been renewed with plasterboard and they are finished with textured finishes of various kinds. The ceiling in the Bathroom is finished with expanded polystyrene tiles.

5.7 The joinery has all been overhauled. At first-floor level on the rear elevation the windows are steel-framed side hung casements. Most of the other windows are wooden, with side hung casements. Internal doors are a mixture but most of them are standard modern doors. The skirting in the lounge is a cement and sand skirting. Other skirtings are of plain modern type on the ground floor and on the first floor they are old square edged boards.

5.8 Domestic fittings are fairly modern. Only one open fireplace remains (in the Dining Room) and this is fairly typical of the late 1940s period. Bathroom and Kitchen fittings are more modern.

SERVICES

6.0 Main water is supplied by the Thames Water Authority and electricity by the Southern Electricity Board. Mains gas is not available in the village. For heating, the house at present relies on point sources of heat. There is an open fireplace and there are also two off-peak electric storage heaters. Mains drainage is connected.

6.1 The property is situated in the Royal County of Berkshire (Newbury District). It is assessed at £202.02, rateable value and the general rates for the current year amount to £276.94. Water and drainage charges are additional and are collected direct by the water supply authority.

EXTERNAL CONDITION

7.0 The main walls are in fair condition for a building of this age and style. They are all reasonably square and plumb, and although many alterations have been made, particularly to the window and door openings, we could see nothing which we could identify as evidence of lack of structural stability. It is unfortunate that the elevations have been painted in the way they have, because this makes it difficult to see much of the fabric of the building. But we can see that the pointing of the brickwork generally is in poor condition, and it was not properly repaired before the elevations were painted. The faces of many of the bricks are soft and perishing. This is fairly typical of the local bricks, and although the paint which has been applied over the surface of the walls will do something to retard the process of decay, there is a problem of dampness in the building. It would appear that damp is penetrating into the fabric of the walls but it is not possible to determine precisely whether it is rising from the ground through lack of an effective damp-proof course or penetrating at the defective pointing and poorly detailed openings. It is important to note that the paint which was applied to keep damp out, will also keep it in. There is very little that can be done to check this decay of the brickwork but probably the best precaution is to render the walls. The green paint is quite out of keeping with the local style of building and there would be much to be said for changing the appearance. It will probably be impossible to strip off the colouring completely, so that even if it can be removed it will still be necessary to apply some other coating. We suggest that the best type of rendering for the purpose would be a textured rendering such as 'Tyrolean' finish, but this particular finish, which is spattered on to the surface, goes on so evenly that it follows every blemish in the brickwork. It will therefore be necessary to prepare the wall by stripping off the paint, making good the pointing and facing up the blown bricks. It will be of little use to do this until the various sources of dampness have been identified and eradicated and the walls allowed to dry out.

7.1 Some of the damp in the main walls is almost certainly due to poor detailing of the window openings and the condition of the sills. The windows on the first floor at the rear have no proper sills at all. Sills do have a particular purpose in keeping out rainwater and therefore should project about 37 mm (1½ in) from the face of the wall. They should have on the underside a groove which will throw off rainwater running underneath the sill, thus preventing the water from running back on to the brickwork below. The sills on the first floor at the front of the house have only a very small projection but this could be extended. On the ground floor some of the sills are affected by wet rot and will need to be replaced. We shall refer again to the condition of these sills and the possible repairs to them later in the report. Of course, if the walls are to be prepared and rendered as we have suggested in the previous paragraph, this must be taken into account when detailing the sills. We very often find problems of this kind arising when windows have been correctly set in the brickwork and then at some time later the walls have been rendered, thus reducing the projection of the sill and sometimes even filling up the groove.

7.2 We examined the covering of the main roof externally from ground level only, with the aid of a pair of binoculars. It is quite obvious that the roof has fairly recently been stripped and re-tiled. The dark brown concrete tiles which have been used are quite different from anything else in the vicinity and out of keeping with local finishes; there is little that can be done but to accept them. There are a few irregularities in the roof line, possible because the tiling was not done very skilfully. It may also be due to the fact that the roof was probably originally slated and that the concrete tiles are heavier than the slates which were previously there. Within the roof void some additional struts had been inserted to support the new roof load and these appeared satisfactory as was the load-bearing internal walls supporting the new struts.

7.3 The chimney head is in fair condition. The mortar fillet where it passes through the roof line has been renewed, presumably at the time when the roof was tiled. This fillet is not as satisfactory as a sheet metal flashing and soaker, although if we were to condemn a mortar fillet in this way we should be condemning many thousands of houses throughout the country where this type of finish is used. The weakness of the junction is that if there is any differential movement between the roof and the stack then the fillet will crack and cease to be weather-proof, whereas a sheet lead detail would be sufficiently flexible to accommodate slight movement. However, we are of the opinion that the roof fillet can be left alone, at least for the time being.

INTERNAL CONDITION

8.0 There is a great deal of damp in the main walls and internal walls. One would expect to find rising damp in a building of this age, where even if there were originally a damp-proof course, it would probably be at the end of its life. We have been told that a damp-proof course has been injected, but no documentary evidence to this effect has so far been produced. By using an electric moisture meter, we found very little evidence of true rising damp in the main external walls. Much of the plaster of these walls is very hard, suggesting that a damp-proof course has indeed been injected and the wall plaster stripped and replaced with a sand and cement rendering. Most of the damp in the outside walls is penetrating damp and this is in the region of the windows. We have suggested above that this is probably due to defective brickwork and poor detailing or the complete lack of window sills. The eaves gutters are badly placed and we think that rainwater running from the roof may overshoot or undershoot the gutters. It was fair weather when we inspected the property and so we were not able to check the gutters in storm conditions. If this is so, they ought to be checked and adjusted or possibly replaced with gutters of a larger section. But we feel reasonably confident that most of the dampness in the outside walls is not true rising damp. There is a further complication because, for most of the length of the outside walls, the surrounding paving is close to and in some cases above the level of the ground floor inside. Therefore, if there is a damp-proof course at or about floor level it may be below the paving level. We would recommend that as far as possible the level of the paving be dropped to 150 mm (6′) below the surface of the internal floors for, not only may the paving be above the damp-proof course, but it is also possible that rainwater falling from the eaves is bouncing back on to the walls above the damp-proof course, thus making it ineffective to some extent.

8.1 There is also some damp in the internal walls, particularly in the wall between the Lounge and the Entrance Hall and also between the Dining Room and the Kitchen, and here there are some wooden skirtings which are rotting. These walls are probably built off the ground, possibly with strip foundations and if we have been correctly informed that a damp-proof course has been inserted in the outside walls, it is possible that the internal walls have been overlooked. Our client has already obtained two estimates for the insertion of a chemical damp-proof course and both refer to re-plastering. The importance of this is that rising damp will bring up from the ground salts which are hygroscopic, that is, they would absorb moisture from the atmosphere so that even if the rising damp itself has been cured, the plaster will continue to attract moisture and will remain damp for some time. It is for this reason that re-plastering is specified. In this case, since we question if all the dampness in the outside walls is true rising damp and since we also think that some re-plastering may have been carried out, we recommend that a chemical analysis be made of samples of the plaster to determine whether or not it is necessary to re-plaster. The specialist damp-proofing contactors are usually able to arrange for this to be done but if they cannot, we can have the laboratory analysis made. In addition to the fungal rot in the skirtings on the ground floor, there is some fungal rot in the old skirtings on the first floor, again signifying penetrating rather than rising damp. These rotten skirtings must be taken out and burnt and replaced by pretreated timbers, with a backing of polythene sheet between the wood and the plaster. The adjoining brickwork and plasterwork should also be chemically treated with a fungicide. There is some damp in the South West flank wall of Bedroom 1, possibly due to a poor detail where the thatched roof of the adjoining house meets it. This will be extremely difficult to deal with, although a silicone injection will give some relief but it will have to be repeated from time to time.

8.2 There is an access hatch to the roof void from the landing and this gives access to the void above the centre and South Western sections of the house. These would originally have been two cottages and the party wall between them does not rise above ceiling level. This was fairly normal in older properties built before about 1880 and is of no great significance in the present case. The tiles are laid over bituminous felt, indicating beyond doubt that the roof has been stripped and re-tiled, as was observed from the outside. The construction of the roof is satisfactory and for its age the timber is in very fair condition. The ridge board and some of the rafters are in contact with the brickwork of the chimney stack in the centre of this part of the building. It is a poor detail and one which would not be permitted today. It can be a fire hazard if the chimney overheats, but there is very little that can be done about it at present, without making a fairly elaborate alteration to the roof structure. It is a condition which is commonly found in old houses. We can only recommend that if the chimney is to be used at all, care should be taken to see that it is kept clean, and if possible, the chimney should be lined.

8.3 There is a second small access hatch from Bedroom 3 and it is within this roof void that we established that this part of the building is a later addition. It has a different type of roof from the original work and the wall between the two sections of roof, and indeed right down through the building, is in facing brickwork, 215 mm (one brick) thick. The roof structure generally is in satisfactory condition and 25 mm (1′) of glassfibre lagging is spread between the joists.

8.4 Both roof voids are unventilated and if, as we recommend later in this report, the insulation is upgraded, condensation within the roof void will occur. This condensation will inevitably lead to timber decay and therefore it is vital that some ventilation be created by introducing air vents at the eaves; these must remain uncovered when the new insulation is laid. We can advise further on this matter if requested. There is a great deal of dirt and dust in all the roof voids and we did not see any evidence of active infestation by beetle, although there are a number of old flight holes. These are old and dirty, but the possibility of continuing infestation cannot be ruled out. In this case, our client has no option as the Building Society has already required specialist treatment of all accessible timber to be carried out.

8.5 The ground floor is in fair condition. The wood mosaic forms quite a pleasant floor finish. In the Dining Room and Kitchen the floor coverings are laid direct on a fine cement and sand screed. We cannot tell whether or not there is any damp-proof membrane in this floor, but since they are obviously quite modern we think it likely that a membrane was incorporated.

8.6 The first floors are all covered. They are in fair condition for their age, but we recommend that they be treated against infestation. Special attention should be given to the edges of the floors to establish whether or not there is fungal rot in the boards or in the ends of the joists, which are probably in contact with damp brickwork. We have already drawn attention to fungal rot in some of the skirtings on the first floor.

8.7 Ceilings have been patched and repaired. They are not particularly elegant, but they appear to have been overhauled and to be in a sound condition.

8.8 The joinery items externally are in a poor condition, particularly the window frames and sills. We detected wet rot in most of the sills to the ground floor windows. The first floor windows do not project adequately from the face of the wall and on the rear elevation where steel windows are fitted, there are no adequate sills at all. We have already referred to the necessity for seeing that sills are properly detailed and maintained. Where modern sills are affected by wet rot, it may be possible to cut them out and replace them, although in the end, the joiner may recommend that it would be less expensive to replace the window frame completely. In the case of the steel casements, it would be possible to cut out a course of brickwork below the window frames and to fit in either a wooden or a precast concrete sill and this is recommended. We have referred above to the areas of skirting which are affected by fungal rot and which need to be replaced. The door and window at the rear of the garage are in poor condition, although this is in effect only an out-house and perhaps the condition can be accepted for the time being.

8.9 Domestic fittings are in fair condition only. The enamel of the bath is chipped in several places. The stainless steel sink is badly stained although it may clean up. The fireplace in the dining room is probably about forty years old and is showing its age.

SERVICES CONDITION

9.0 Water enters the house through a stop valve located in the garage. All the plumbing we can see is in modern copper tube. The valve itself is stiff and it should be freed to make sure that it can be turned off without difficulty. The rising main supplies a plastic cold water cistern located in the linen cupboard on the first floor which should be covered to keep out dirt and dust. Hot water is provided by an electric immersion heater in the copper hot water cylinder in the same linen cupboard. This hot water cylinder has an insulating jacket which is in poor condition. Taps run satisfactorily, but most of them need new washers. Therefore, some minor repairs are needed to the water installation.

9.1 Electricity is supplied through meters in the cupboard in the Entrance Hall. The off-peak circuit has a credit meter but the ordinary circuit has a pre-payment meter taking '50 pence' pieces. Some of the wiring is certainly new, but within the roof void we found some old tough rubber sheathed cable and some repairs are needed. The immersion heater is connected by tough rubber sheathed cable to a socket rather than by heat-resisting cable to a spur outlet. We recommend that the electrical installation be tested and checked by a competent electrician and repaired as found necessary. We can arrange for this to be done when instructed.

9.2 For heating, the house at present relies on point sources of heat. There is 25 mm $(1 \leq)$ of glassfibre lagging in the roof void. This is less than modern performance standards require and should be improved for economic reasons. This lagging will have to be taken out for the timber to be treated against beetle and it will probably then not be reusable.

9.3 Foul drainage is taken to the main sewer. There is one inspection chamber near to the North Eastern end of the rear elevation. We lifted this cover and the underground works appear to be fairly modern and in satisfactory condition, but the drains should be rodded and flushed through to make sure that they are clean. A wire balloon should be fitted at the head of the vent pipe to keep out the birds.

9.4 Roof water is collected at the eaves in modern half round gutters of grey PVC and brought down in discharge pipes of the same material. The downpipe at the front of the house is finished well above the level of the paving and is at present picked up in a section of half round gutter and taken away clear of the paving. The downpipe at the side goes underground to a soakaway. The eaves gutters are not very satisfactory as their alignment is suspect and we are of the opinion that they have been incorrectly placed under the eaves. There is some staining in the upper walls suggesting that the discharge from the roof either overshoots or undershoots the gutters. They need to be checked in heavy rain to see exactly how they are performing. Evidence suggests that some repairs and adjustments are needed. As to the discharge over the ground, while it is not the best practice, it is quite common and generally acceptable in rural districts, provided that the rainwater is not allowed to pond against the lower walls. We have already recommended that the paving be dropped in level and when it is re-laid, if it is concrete or slab paving, then it should be laid to a slight fall away from the house. The downpipes could then be brought down almost to the level of the paving and channel contructed in the paving to take the discharge away clear of the building. We think that the condition of the gutters and downpipes is at least partly responsible for the amount of dampness which is clearly penetrating through the walls of the building.

EXTERNAL WORKS

10.0 There is a modest garden, approximately level and nearly rectangular. The solicitor should check the location of the boundaries and advise as to any responsibility for the erection and maintenance of fences. The road is fully made up and is maintained by the Berkshire County Council.

SUMMARY AND CONCLUSIONS

11.0 This is a pleasant enough house in its way, formed by the conversion of three small cottages. Someone in the past has carried out a fair amount of work to it but it is our opinion that the building as it now stands is something of a monument to the inadequacies of 'do-it-yourself'. The roof has been re-tiled, not with the most suitable of materials but evidently adequately. Undoubtedly the biggest problem in the building is the damp penetration and we think that most of this is penetrating rather than rising damp.

11.1 In the course of our report we have drawn attention to the following items:

    (a)   The solicitor should check very carefully on the title and the boundaries.

    (b)   The pointing of the brickwork is generally in poor condition.

    (c)   There is a fairly large number of bricks which are perishing or partly perished.

    (d)   Consideration should be given to making good and rendering the elevations, but proper steps should be taken to make sure that the walls dry out, as far as possible, before rendering. In addition, the window sills need to be properly detailed to take account of any new rendering.

    (e)   We recommend that the paving round the house be dropped to 150 mm (6') below ground floor level.

    (f)   We recommend that samples of the wall plaster be taken for analysis to determine whether or not they are affected by hygroscopic salts.

    (g)   We recommend that all accessible timbers of the roof and wooden floors be treated against beetle infestation.

    (h)   We recommend that the insulation to both roof voids be improved and that ventilation is introduced via eaves vents.

    (i)   The first floor, particularly at the edges, should be checked for traces of fungal rot and this can be done at the same time as the treatment for woodworm is carried out.

    (j)   Some repairs and replacements are needed to wooden window sills and skirtings and the window openings, which have no sills, should be provided with them.

    (k)   Some minor repairs are needed to the water installation.

    (l)   We recommend that the electrical installation be tested and repaired as found necessary.

    (m) We recommend that the drains be rodded and flushed through and a wire balloon should be fitted at the head of the vent pipe.

    (n)   We recommend that the rainwater gutters be checked carefully in wet weather, to see exactly how they are performing, since they are at least partly responsible for damp inside the building.

11.2 We would estimate the costs of investigation and repair as follows:

    (a)   Repointing and replacing perished brickwork (items 11.1 b & c, refers) £350.

    (b)   Making good to rendering (item 11.1 d, refers) £250.

    (c)   Dropping paving around the property (item 11.1 e, refers) £150.

(d) Investigating the plaster for salts and assessing the extent of rising damp and fungal infection (items 11.1 f & i, refers) £225.

(e) Insertion of damp-proof course including re-plastering internally £850.

(f) Treating all timber for woodworm (item 11.1 i, refers) £400.

(g) Improving detail around window openings (item 11.1 d & j, refers) including repairs and replacement of sills and skirtings £450.

(h) Improving insulation to both roof voids (item 11.1 h, refers) £250.

(i) Minor repairs to water installation (item 11.1 k, refers) £100.

(j) Rodding and flushing drains (item 11.1 m, refers) £150.

(k) Inspecting and realigning gutters, as necessary (item 11.1 n, refers) £75.

All these costs are approximate and do not include VAT.

11.3 As to the order of priority in which this work needs to be carried out, it is quite clear that the most urgent matter is to eliminate the damp from the building. Since there does not appear to be any documentation relating to a damp-proof course, it is almost certainly necessary to insert one and to have this done with the benefit of the appropriate guarantees. We think that much of the damp in the building is penetrating from the defective window-opening details and possibly from defective rainwater gutters. Once the penetration of damp has been checked it will take some months for the walls to dry out. There is no reason why this should delay the treatment of timber against infestation and the repairs to the services. When this has been carried out, redecoration can be put in hand, but it will be of little use to re-decorate the walls which are damp. At present, the dampness in the house shows most clearly in the form of stained decorations. It may become necessary to live for a year or so with the interior of the house in need of decoration. The repair and rendering of the outside walls ought to have a fairly high priority because of the importance of maintaining the weather-proof condition of the walls, but the other sources of penetrating damp should be eliminated first.

A SURVEYOR, ARICS, MCIOB,   24th APRIL 1986
SURVEYING SERVICES,
HAWTHORN COTTAGE,
WINTERBOURNE,
NEWBURY,
BERKSHIRE

Appendix D: Period Classification

In the early stages of civilisation, man learnt to use indigenous materials to construct temporary shelters from which he could carry out short-term hunting expeditions. Later he formed communities and learnt to build more permanent shelters and barricades against marauding bands. He looked for his materials in their natural forms such as timber, stones and clays but in later periods he began to process materials to make them more adaptable to building for particular purposes. By mixing primary materials and applying heat he found that he had a new material which could be shaped easily and which would set hard when mixed with water. He also found that the mixture could be burnt into very hard and durable units. In most cases these materials were on or near the surface so they were obtainable wherever strata outcropped at the surface. Man was able to identify the limits of these materials and as a result many communities were set up near to these supplies. In the later years he learnt to quarry and mine the materials to save trekking long distances.

The communities became noted for their particular material source and a trade of sorts was developed to exchange materials, although movement overland was extremely difficult. However, movement was easier on the water and coastal trade developed where the materials were found near to the coast.

The strength of the community buildings was centred around stone, bulk timber and a form of mortar which gave rigidity, but it was much later that such mortar developed into a masonry-bonding medium and allowed the building of thinner walls.

Communities also developed skills in processing and constructing with the materials and these skills were developed and exported to other communities. Even today, we see evidence still of these skills being centred around certain areas in the UK, for example, the more exacting skills, such as those requiring controlled heat, can be found in areas of coal-bearing strata. The same influence affected skills requiring large quantities of water in the processing. The developing skills in building soon polarised in certain areas where the material was abundant or easily won, but as the demand for processed materials developed, centres of manufacture were set up either in areas where the main ingredient was easily obtainable or in the fringe area between two of the ingredients. This was particularly noticeable in the quarrying of stone and in the manufacture of burnt clay products. In both these industries it was unnecessary to have any elaborate plant, unlike that of the manufacture of iron and steel when the plant needed required a high initial capital outlay.

Because of the local deposits of the raw materials and the intensive labour applied to these materials, skills of the workmen developed to a high degree and their reputation spread, but generally the skills were maintained at a useful productive level. The workman looked to the amount of work he could do, the demand for and the value of his skill. If he became too skilful the demand became sporadic and payment irregular, so it can be seen that local industry was for the most part related directly to the needs of the community, especially in building. Carpenters and masons were normally in constant demand for a basic building structure and although some became well-known figures in their craft, the majority worked to a design suggested by the client. Once their work had been completed they would seldom be called upon again for some considerable time because of the durability of the materials they used. In the case of the thatcher, plasterer and the decorator, the material had a much shorter life-span and often these tradesmen would move around the district doing their work again and again.

As the network of roadways and other communication links developed, the processing and the constructing skills separated and the processing became centred within a large distribution area. The materials were carried long distances by water, road and rail and they were bound only by the economics of trading. More and more processing was carried out at the base and gradually the development of components began to take precedence over raw materials in some areas. Slate was riven, cut, holed and packaged at base. Stones were dressed, shaped to specific details, batched and transported. Items of clayware were manufactured and given a final finish which left only the fixing to complete the work. Gradually, the whole process of acquiring the raw material, processing and fixing it has changed from a local industry to a nationwide practice. It is not unusual now to see almost complete buildings being transported from factories to sites and even this may be taken to the extreme of demountability whereby the client can choose where he wants his building for any particular length of time.

The history of buildings can be traced to some degree by various means and perhaps the obvious start in this enquiry would be talking to those who are associated with it. However, in many cases there will be documentary evidence available but even a casual inspection of the physical appearance and planning may give important information which can help to fix a date fairly precisely. It may not be possible, of course, to date a building with great accuracy simply by using architectural evidence and undoubtedly there are dangers in relying solely on one source of evidence. Consequently a knowledge of architectural styles is very important and this should be developed by reading and more especially by observation. *The Illustrated Handbook of Vernacular Architecture* by R. W. Brunskill (published by Faber & Faber) and *Your House—the outside view* by John Prizeman (published by Quiller Press) are particularly good texts on this subject.

In earlier days, materials for walls and roofs would have been obtained locally as previously mentioned, and so one can investigate buildings of some age on that basis. Materials that were not obtained locally can indicate either a rich owner when the building was developed or improvement in communication and transport. For example, river communications improved at the beginning of the eighteenth century, canal navigation about 1775 and the explosion of improved transport by road from about 1840. Roads had been improving over a considerable period, but the weight and size of materials and the lack of roads to areas where materials were found naturally made the contribution of the road to the dispersal and more widespread use of materials not so significant. Windows and doors, if original, can also be evidence of the period, and these elements should be carefully observed.

If a documentary investigation is to be undertaken, the Ordnance Survey maps may be the first approach. The first maps dated from the early part of the nineteenth century and by 1840 the whole of the south of England was mapped at a scale of one inch to a mile. Many towns in Britain were mapped from 1850 and these can be most useful in urban areas. Tithe maps may be useful, as are deeds and related documents if precise dating is necessary, but it is seldom that great accuracy is required.

Those buildings of comparatively recent date do not present difficulty and documentary evidence is usually available. Clues may be found in the names of roads and districts and a knowledge of these and development of land, the changing appearance of the countryside and 'local' experience will help to deduce the information required. Buildings that are 'listed' will often provide useful information on building styles and materials locally.

Listed buildings

The definition of a listed building is wide and it can include structures other than **buildings**. If the 'listing' procedure has been applied, the criteria laid down will have been **satisfied** and a category allocated. Details are given in the legislation and various publications but generally the criteria are:

(a) Buildings erected before 1700 and remaining in a condition approaching original.
(b) Buildings between 1700 and 1840 are likely to be selected and listed.
(c) Selected buildings (quality, character, work of principal designers, forming part of a special group or of special technological interest) between 1840 and 1914.
(d) Buildings constructed between 1914 and 1939 are presently being selected and listed.

The classification is in ranking of importance and is defined as

Grade I   buildings of exceptional interest.
Grade II   buildings of special interest which attract effort to preserve them and some of the more important will be grouped in Grade II*.

A copy of the listings is available from the appropriate authority and no work of demolition, alteration or extension which affects the original character can be carried out without listed building consent, and contravention may be punished by a fine, imprisonment or both. Rights of entry are provided under several sources for officers and officials to carry out duties connected with a listed building, for example, in connection with the serving of a repairs notice. It is important therefore to carry out a thorough investigation in the office if the inspection is to be of a building known to be listed or which one suspects may be listed.

Bibliography

Building Research Station, Garston, Watford WD2 7JR.

BRE *Digests 45* and *46*: 'Design and Appearance'.

BRE *Digest 54*: 'Damp-proofing solid floors'.

BRE *Digests 63, 64,* and *67*: 'Soils and foundations Parts 1, 2 and 3'.

BRE *Digest 73*: 'Prevention of decay in window joinery'.

BRE *Digest 77*: 'Damp-proof courses'.

BRE *Digest 89*: 'Sulphate attack on brickwork'.

BRE *Digest 98*: 'Durability of metals in natural waters'.

BRE *Digest 110*: 'Condensation'.

BRE *Digest 160*: 'Mortars for brickwork'.

BRE *Digest 176*: 'Failure patterns and implications'.

BRE *Digest 180*: 'Condensation in roofs'.

BRE *Digest 196*: 'External rendered finishes'.

BRE *Digest 201*: 'Wood preservatives : pretreatment application methods'.

BRE *Digest 218*: 'Cavity barriers and ventilation in flat and low-pitched roofs'.

BRE *Digest 236*: 'Cavity insulation'.

BRE *Digests 240, 241* and *242*: 'Low-rise buildings on shrinkable clay soils, Parts 1, 2 and 3'.

BRE *Digest 245*: 'Rising damp in walls : diagnosis and treatment'.

BRE *Digest 251*: 'Assessment of damage in low-rise buildings'.

BRE *Digest 257*: 'Installation of wall ties in existing construction'.

BRE *Digest 261*: 'Painting woodwork'.

BRE *Digest 262*: 'Selection of windows by performance'.

BRE *Digest 263*: 'Part 1 Mechanism of protection and corrosion'.

BRE *Digest 264*: 'Part 2 Diagnosis and assessment of corrosion-cracked concrete'.

BRE *Digest 265*: 'Part 3 The repair of reinforced concrete'.

BRE *Digest 266*: 'Sound insulation of party floors'.

BRE *Digest 268*: 'Common defects in low-rise traditional housing'.

173

BRE *Digest 270*: 'Condensation in insulated domestic roofs'.
BRE *Digest 280*: 'Cleaning external surfaces of buildings'.
BRE *Digest 293*: 'Improving the sound insulation of separating-walls and floors'.
BRE *Digest 296*: 'Timbers : their natural durability and resistance to preservative treatment'.
BRE *Digest 297*: 'Surface condensation and mould growth in traditionally-built dwellings'.
BRE *Digest 298*: 'The influence of trees on house foundations in clay soils'.
BRE *Digest 304*: 'Preventing decay in external joinery'.
BRE *Digest 307*: 'Identifying damage by wood-boring insects'.
BRE *Report* (1987): 'Recognising wood rot and insect damage in buildings'.

Information Paper IP 28/79: 'Corrosion of steel wall ties : recognition, assessment and appropriate action'.
Information Paper IP 10/80: 'Avoiding joinery decay by design'.
Information Paper IP 2/81: 'The selection of doors and doorsets by performance'.
Information Paper IP 4/81: 'The performance of cavity wall ties'.
Information Paper IP 16/81: 'The weather stripping of windows and doors'.
Information Paper IP 21/81: 'In-situ treatment for existing window joinery'.
Information Paper IP 12/82: 'House Longhorn Beetle Survey'.
Information Paper IP 7/83: 'Window to wall jointing'.
Information Paper IP 4/84: 'Performance specifications for wall ties'.
Information Paper IP 10/84: 'The structural condition of some prefabricated reinforced concrete houses designed before 1960'.

Technical Note 44: 'Decay in buildings : recognition, prevention and cure'.

Current Paper CP 6/76: 'Results of field tests on the natural durability of timber (1932–1975)'.
Information Sheet 15 20/77: 'Painted weathered timber'.

Timber Research and Development Association (TRADA)
Information sheet No 3: 'Introduction to timber framed housing'.
Information sheet No 5: 'Timber framed housing – specification notes'.
Information sheet No 10: 'Structural surveys of timber frame houses'.

Royal Institution of Chartered Surveyors (RICS)
Practice note 'Structural Surveys of Residential Property', 2nd edn, 1985.
House Buyers Report and Valuation, 3rd edn, 1987.
Flat Buyers Report and Valuation, 1983.
Building Surveyors Guidance Note 'Dilapidations', 1983

British Standards Institution
BS 402: 1979 'Clay plain roof tiles and fittings. Part 2 : 1970'.
BS 473 and *BS 550: 1971* 'Concrete roofing tiles and fittings'.
BS 1202, parts 1, 2 and 3: 1962.
BS 5617: 1985 'Specification for urea–formaldehyde (UF) foam systems suitable for thermal insulation of cavity walls with masonry or concrete inner and outer leaves'.

BS 5618: 1985 'Code of practice for thermal insulation of cavity w\5
or concrete inner and outer leaves) by filling with urea–formalde
systems'.

BS 8208: 'Guide to assessment of suitability of external cavity walls fo
thermal insulants'.

BS Code of Practice 144: 'Roof coverings Part 4: 1970 Mastic asphalt'.

BS Code of Practice 144: 'Roof coverings Part 3: 1970 Built-up bitumen felt'.

BS Code of Practice, C.P. 2001: 1957, 'Site investigations'.

Maintenance Management – a guide to good practice, 2nd edn, 1982, published by
Chartered Institute of Building.

Lyall Addleson, *Building Failures – A Guide to Diagnosis, Remedy and Prevention*,
published by the Architectural Press.

Handbook and Directory of Members, published by the Stone Federation, 82
Cavendish Street, London, W1M 8AD.

Department of the Environment, *Housing Defects Act 1984, The Housing De
(Prefabricated Reinforced Concrete Dwellings) (England and Wales) Designations
Supplementary Information*.

H.W. Fowler, *A Dictionary of Modern English Usage*, 2nd edn, 1982, published
Oxford University Press.

E. Gower, *The Complete Plain Words*, 3rd edn, revised by Sidney Greenbaum and J
Whitcut, 1986, published by HMSO.

The Building Regulations 1985, published by HMSO.

Structural Survey, published quarterly by Henry Stewart Publications, 88 Baker Str
London W1M 1DL.

Bats In Roofs: A Guide for Surveyors, published by The Nature Conservancy Council.

R.W. Brunskill, *The Illustrated Handbook of Vernacular Architecture, 1986*, publishd
by Faber and Faber.

John Prizeman, *Your House – the outside view*, 1982, published by Quiller Press.

Ivor H. Seeley, *Building Surveys, Reports and Dilapidations*, published by Macmillan
Education.

BS 5618: 1985 'Code of practice for thermal insulation of cavity walls (with masonry or concrete inner and outer leaves) by filling with urea–formaldehyde (UF) foam systems'.

BS 8208: 'Guide to assessment of suitability of external cavity walls for filling with thermal insulants'.

BS Code of Practice 144: 'Roof coverings Part 4: 1970 Mastic asphalt'.

BS Code of Practice 144: 'Roof coverings Part 3: 1970 Built-up bitumen felt'.

BS Code of Practice, C.P. 2001: 1957, 'Site investigations'.

Maintenance Management – a guide to good practice, 2nd edn, 1982, published by the Chartered Institute of Building.

Lyall Addleson, *Building Failures – A Guide to Diagnosis, Remedy and Prevention*, 1982, published by the Architectural Press.

Handbook and Directory of Members, published by the Stone Federation, 82 New Cavendish Street, London, W1M 8AD.

Department of the Environment, *Housing Defects Act 1984, The Housing Defects (Prefabricated Reinforced Concrete Dwellings) (England and Wales) Designations 1984, Supplementary Information*.

H.W. Fowler, *A Dictionary of Modern English Usage*, 2nd edn, 1982, published by Oxford University Press.

E. Gower, *The Complete Plain Words*, 3rd edn, revised by Sidney Greenbaum and Janet Whitcut, 1986, published by HMSO.

The Building Regulations 1985, published by HMSO.

Structural Survey, published quarterly by Henry Stewart Publications, 88 Baker Street, London W1M 1DL.

Bats In Roofs: A Guide for Surveyors, published by The Nature Conservancy Council.

R. W. Brunskill, *The Illustrated Handbook of Vernacular Architecture, 1986*, published by Faber and Faber.

John Prizeman, *Your House – the outside view*, 1982, published by Quiller Press.

Ivor H. Seeley, *Building Surveys, Reports and Dilapidations*, published by Macmillan Education.

Index